Energy Law and Policy in a Climate-Constrained World

Energy Law and Policy in a Climate-Constrained World

Victor Byers Flatt
Alfonso López de la Osa Escribano
Aubin Nzaou-Kongo

Westphalia Press
An Imprint of the Policy Studies Organization
Washington, DC
2022

ENERGY LAW AND POLICY IN A
CLIMATE-CONSTRAINED WORLD
All Rights Reserved © 2022 by Policy Studies Organization

Westphalia Press
An imprint of Policy Studies Organization
1367 Connecticut Avenue NW
Washington, D.C. 20036
info@ipsonet.org

Library of Congress Control Number: 2022916601

ISBN: 978-1-63723-915-5

Cover and interior design by Jeffrey Barnes
jbarnesbook.design

Daniel Gutierrez-Sandoval, Executive Director
PSO and Westphalia Press

Updated material and comments on this edition
can be found at the Westphalia Press website:
www.westphaliapress.org

Table of Contents

List of Abbreviations ix

List of Contributors xii

Acknowledgments xv

Introduction xvii

1. The Europeanisation of Renewable Energy Consumption Targets Renewed by Directive (EU) 2018/2001 1

1.1 Introduction 1

1.2 The Level 4

    1.2.1 Increase in the Share of Renewable Energy in Overall Final Energy Consumption 5

        1.2.1.1 Energy Policy Choices 5

        1.2.1.2 Climate Policy Choices 6

    1.2.2 Increase in the Share of Renewable Energy in the National Energy Mix 10

        1.2.2.1 Europeanisation of Part of the National Energy Mix 11

        1.2.2.2 Partial Europeanisation of the National Energy Mix 12

1.3 The Scope 13

    1.3.1 Overall Objective 13

        1.3.1.1 Including a Sectoral Target 13

        1.3.1.2 Does Not Include National Target 15

    1.3.2 Binding Target 18

        1.3.2.1 Collective Obligation of Result 18

        1.3.2.2 Partial Europeanisation of Resources 19

1.4 Conclusion 21

2. External Relations of the European Union in the Energy Sector: EU-Russia Energy (Trade) Relations 23

2.1 Introduction 23

2.2 EU Energy Consumption and Production, Current Trends and Challenges 24

2.2.1 EU Consumption 24

2.2.2 EU Production and International Trade Supply 26

2.2.3 Trends and Challenges in the EU External Relations in the Energy Sector 27

2.3 Competence Issues Regarding the EU Treaties 29

2.4 Participation in Treaty Regimes 31

2.5 Participation in the Energy Community 34

2.6 EU-Russia Trade Relations in the Energy Sector 37

2.7 Conclusion 40

3. THE EUROPEAN UNION AND THE LEGAL GOVERNANCE OF OFFSHORE ENERGY ACTIVITIES 45

3.1 Introduction 45

3.2 From Cooperation in Public International Law 49

3.2.1 In a Material and/or Formal Soft Law Framework 49

3.2.1.1 Regional Systems where the EU is not a Party 49

3.2.1.2 When the UE is a Party to Regional Treaties 56

3.2.2 In a Hard Law Framework: the Exception of the Mediterranean 61

3.2.2.1 A Protocol Dedicated to Offshore Activities 56

3.2.2.2 The EU being a Party to the Protocol 67

3.3 To Integration into European Union Law 70

3.3.1 The European Union Law Acquis 70

3.3.1.1 Integration of the Mediterranean Acquis 71

3.3.1.2 For EU Member States 73

3.3.2 The Contribution of Directive 2013/30/EU 79

3.3.2.1 The Directive on Safety of Offshore Oil and

Table of Contents

Gas Operations 80
3.3.2.2 Challenges for EU Blue Growth 80
3.4 Conclusion 89

4. THE PLAYERS IN THE NEW ENERGY SYSTEM: WHAT ROLE FOR THE STATE IN THE ANTHROPOCENE ERA? 95
4.1 Introduction 95
4.2 The Role of the State: An Old Question Asked Anew 100
4.2.1 A Constitutional Law Rationale 100
4.2.2 An Energy Law Perspective 105
4.2.2.1 Statutory Arrangements to Energy Transition 105
4.2.2.2 Policy Objectives for Energy Transition 109
4.3 Towards an Effective Transition Effort? 112
4.3.1 The Role of Steering the Transition 112
4.3.2 The Role of Fostering the Energy Justice 116
4.4 Energy Transition: Between International Cooperation and Global Dynamics 122
4.4.1 A Contribution Within a Global Dynamic 123
4.4.2 The Specific Situation of Developing Countries 127
4.5 The African Union's Efforts Towards the States 132
4.6 Conclusion 136

Epilogue 141
Index 149

LIST OF ABBREVIATIONS

ACEC Africa Clean Energy Corridor

AREI African Renewable Energy Initiative

CCS carbon capture and storage

CDM Clean Development Mechanism

CJEU Court of Justice of the European Union

CO_2 Carbon Dioxide

COP Conference of the Parties

CMP Conference of the Parties serving as the Meeting of the Parties to the Kyoto Protocol

COP Conference of the Parties to the Kyoto Protocol

CSD Commission on Sustainable Development (UN)

CTF Clean Technology Fund

CUA Commission of the African Union

DDP Deep Decarbonization Pathways

EC European Community

EC European Commission

ECT Energy Charter treaty

EnC Energy Community

EnC Treaty Energy Community Treaty

EU   European Union

EURATOM European Atomic Energy Community

GA general assembly

GATT General Agreement on Tariffs and Trade

GHG greenhouse gas

GDP gross domestic product

GSI sustainability indicators for bioenergy

IA international agreements

ICJ International Court of Justice

IDU Independent Delivery Unit

IEA International Energy Agency

IEL international environmental law

IMO International Maritime Organization

INECP integrated national energy-climate plan

IPCC Intergovernmental Panel on Climate Change

IRENA International Renewable Energy Agency

ISA the International Solar Alliance

ISDS investor-state dispute settlement

Lisbon Treaty of Lisbon

MARENA Mauritius Renewable Energy Agency

MEA multilateral environmental agreement

MDGs Millennium Development Goals

MEPS Guidelines for Minimum Energy Performance Standards

MOP Meeting of the Parties

NATO North Atlantic Treaty Organisation

OECD Organisation for Economic Cooperation and Development

OJ Official Journal

OPEC Organization of Petroleum Exporting Countries

Abbreviations

RC Republic of Congo

REs Renewable Energies

REDD reducing emissions from deforestation and degradation

SD sustainable development

SDGs sustainable development goals

SE4ALL Sustainable Energy for All

STC specialized technical committees

STC-TTIIET Specialized technical committees on Trade, Industry and Minerals

TEC Treaty establishing the European Community

TEU Treaty on European Union

TFEU Treaty on the Functioning of the European Union

TFM Technology Facilitation Mechanism

UK United Kingdom

UN United Nations

UNCLOS United Nations Convention on the Law of the Sea

UNCTAD United Nations Conference on Trade and Development

UNDP United Nations Development Programme

UNEP United Nations Environment Programme

UNESCO United Nations Educational, Scientific and Cultural Organization

UNFCCC United Nations Framework Convention on Climate Change

WACEC West Africa Clean Energy Corridor

WMO World Meteorological Organization

WTO World Trade Organization

List of Contributors

Iuliia Dragunova, Queen Mary University of London.

Victor B. Flatt is the Dwight Olds Professor of Law and is the Faculty Co-Director of the Environment, Energy, and Natural Resources (EENR) Center at the University of Houston Law Center.

Bernadette Le Baut-Ferrarese is Professor of Public Law at University of Lyon 3 Jean Moulin.

Rafael Leal-Arcas is a Jean Monnet Chaired Professor in EU International Economic Law and Professor of Law, Queen Mary University of London (Centre for Commercial Law Studies), United Kingdom. Visiting Professor, New York University Abu Dhabi (United Arab Emirates). Member, Madrid Bar. Ph.D., European University Institute; M.Res., European University Institute; J.S.M., Stanford Law School; LL.M., Columbia Law School; M.Phil., London School of Economics and Political Science; J.D., Granada University; B.A., Granada University.

Alfonso López de la Osa Escribano is the Director of the Center of U.S. and Mexican Law, and Adjunct Faculty on "Comparative Health Law" of the Health Law and Policy Institute at the University of Houston Law Center.

Aubin Nzaou-Kongo is a Marie Sklodowska-Curie Fellow in Law and Energy Policy Researcher at the Environment, Energy, and Natural Resources (EENR) Center, and at the Center of U.S. and Mexican Law at the University of Houston Law Center (U.S.). He is a MSCA Fellow at the Centre for European Studies (CEE), and a member of the Research Group on Comparative, European, and International Law (EDIEC), and an Assistant Professor of Law at the University of Lyon 3 (France).

Nathalie Ros is Professor of Public International Law, University of Tours (IRJI), and Vice-President and Secretary-General of the International Association of the Law of the Sea.

Acknowledgments

This book is a contribution of various scholars to the ever-changing and evolving field of energy law and policy. The wealth of legislative, regulatory, and jurisprudential developments brings a new dynamic to this discipline every day. Therefore, apart from the research object, which is common to the contributors of this book, each doctrinal voice expressed remains unique. To achieve this uniqueness, research needs to be supported and funded. One of the organizations that daily invests in supporting research on energy transition in the multiplicity of forms it may take is the European Commission. In this respect, we are fully grateful to the European Commission for funding the publication of this book. In order to meet the publicity and disclosure requirements for research projects supported by the European Commission, we make the following acknowledgement.

> The Project "NGOs & Transition Governance in Law" (TGL) is a research project funded under the European Union's Horizon 2020 research and innovation program, under the Marie Skłodowska-Curie grant agreement n° 845118.

We would also like to thank Westphalia Press (our publisher), Daniel Gutierrez in particular, Executive Director, and his team, for their enthusiastic acceptance of the project when it was submitted to them, but also for their accurate and rigorous proofreading. It seems appropriate to express our gratitude for their support and patience throughout this process.

At the origin of this collective effort was a larger number of authors. For various reasons, only a few authors have kept this commitment. Our thanks go to them, for they believed in this project and in its outcome.

It is also our duty to thank many colleagues in various institutions without whom this work would not have been possible.

First of all, our colleagues at the University of Houston Law Center. The work conducted under this project is made possible by the daily efforts of frontline workers. Among them are several people in the Dean Law and IT departments. They include Ed Jones, Jacqueline Flores, Ruth

McCleskey, Jaymie Turner, Baroness Adams, Hawthorne Elena, Tanisha Green and Charlette Jefferson. Then, there are all our colleagues at the Environment, Energy and Natural Resources (EENR) Center, who will allow special mention to be made of Professors Gina Warren, and Tracy Hester. Nor can we forget all our colleagues at the Center for U.S. and Mexican Law for their relentless support, especially Celeste Elias, Felix Gomez Umanzor, and Jamie Zarate.

Our thanks inevitably go to our colleagues at the University of Lyon 3 Jean Moulin. The work carried out within the framework of this project would not have been possible without the considerable help of Véronique Gervasoni. From the outset, she has been managing and carrying forward the work for us at the Equipe de Droit International Européen et Comparé (EDIEC). Finally, it is thanks to Julien Roussel and Soline Beaud, from the Research Department, that we were able to coordinate information and action with the Commission's teams.

We would like to express our deepest gratitude to all of them.

Introduction

Victor B. Flatt, Alfonso López de la Osa Escribano and Aubin Nzaou-Kongo

Energy law and policy is in transition, the scope of which is wide and vast. This discipline itself is accompanying a profound social, environmental, and economic change or transformation. One could not diminish the significance of aspects of the transition that are not apparent in the overall picture. This book addresses different aspects of the phenomenon and tackles the energy field from the double perspective of law and policy. Scholars from France, the U.K., and the U.S. have written chapters on different areas of energy law and policy, analyzing different instruments, provisions, and objectives, and questioning the role of actors and institutions. This perspective stems from a broader view of the energy transition and security, but mostly from different approaches relying on law, politics, science, etc. This book presents reflections on concepts, foreign policy, regional and international cooperation, and the specific role the state is to play when it comes to such thing as energy law and policy.

Energy is the preeminent concern of the 21st century—problem of energy for daily survival—coupled with the problem of survival of life on earth. The question of energy carries with it the very critical aspect of how to sustain life on earth now and in the future, knowing that we have compromised the chances of maintaining the natural balances favorable to the survival of human life. One way to look at it is as if it were a vague sketch or an uncertain outline of a much larger problem. In any case, it poses a fundamental question for the survival, struggle, and sustainability of the lives of billions of people on earth. One way of framing the question is, for example, how do you make a hospital system, which in normal times is energy intensive, work during a pandemic? The question arises because the lives of millions of people depend on the availability of ventilators and other essential services. In these circumstances, a power outage is all that is needed to increase the death toll due to the problem of access to energy. The real count of which may be difficult to determine depending on the region of the world where one lives.

At the time of writing, it is impossible to foresee or predict the consequences of the profound correlation between the energy access problem and the death toll of the current COVID-19 pandemic. Of course, no one can predict the aftermath of the ongoing crisis. While there is a general assumption that the death toll is due to the offensive nature of the virus itself, for people who have lived in areas where the energy burden has undermined the health care system and decimated lives, it will be necessary to look at this issue with the benefit of hindsight at some point.

This example shows that the availability of energy in the short and long term is not only a question of future development, but also a question of survival today. Our daily lives depend crucially on it now, while the energy mix does not necessarily meet our daily needs. But it is also here that the question of the energy crisis inevitably resides, notably the idea that a fossil fuel-based mix that has difficulty in meeting current and future needs is confined to a double constraint. On the one hand, the pressure of climate and natural constraints, which means that the energy system is now almost naturally on the verge of collapse as soon as a major event occurs and threatens to break even the optimal functioning of the system. On the other hand, the pressure due to health constraints, which means that the continuity of supply and guaranteed access to energy can promote the availability of energy to manufacture the means of prevention and control of disease, including available treatments, vaccines, masks or face shields. But also for the operation of hospitals for an adequate response to the most serious cases, and for the less serious cases treated at home where the need for oxygen is also expressed. In both cases, we have the illustration in times of crisis of the difficulty for energy to provide essential services when daily survival deeply depends on it. However, it is almost impossible to precisely identify the situations of daily survival, in "normal times," where energy should meet the needs, without indeed imposing arbitrary choices that would exclude even the most critical situations.

The question then—one that is addressed by energy law and policy—is how primary energy sources, both nonrenewable and renewable, can contribute to an energy mix, globally, regionally, nationally, or locally, to ensure access to the essential services provided by energy and to maintain appropriate continuity of energy supply, so that disruptions that may occur do not collapse the entire energy, social, and economic system? The challenge facing energy law and policy is not to limit itself to the primary positivism of presenting the relevant provisions of laws and regulations, but rather to reinforce this approach with a critical perspective that correlates the existing laws with the elements of practice, because energy is undoubtedly the most practical thing. Such a practice can be articulated around the above-mentioned examples of everyday survival, but also how this factor is considered in the development of energy laws and policies. Do we need an energy transition that is the product of reality on the ground, including daily life, or that of technocrats sitting on piles of paper in an office in Mexico City, Brazzaville, Madrid, Singapore, Dubai, Sao Paulo, etc.? The challenge is also for energy law and policy to integrate various factors into the analysis, including the economic, health, social and environmental costs, benefits, and risks of all these energy sources, in order to make their joint contribution to the overall analysis of the new energy system more critical. Above all, it would be interesting to justify based on these elements what policy and legal orientations and priorities could best support a true transition based on the contribution of renewable and fossil energies, while phasing out the latter.

It is safe to say that the development of energy law and policy is consubstantially linked to that of environmental law and policy. In this respect, historically both disciplines have integrated the need to achieve a coherent scheme on securing human energy needs, achieving energy efficiency and energy conservation efforts, while managing to integrate health and safety risks related to energy sources. Bearing in mind the protection of both life and the atmosphere. The contribution of scholars in this transitional phase is more than important at the international, regional, national, and local levels. In each of these spheres, a body of energy law and policy has developed that runs through most of the chapters in this book.

The chapters allow for developments that address in some detail the key concepts and classic and new fields of energy law and policy.

Chapter 1 focuses on how Directive (EU) 2018/2001 aims to raise the EU's level of ambition with regards to renewable energy consumption targets. In this regard, it emphasizes that increasing the share of renewable energy in overall final energy consumption is primarily a matter of public policy. However, a central justification of this public policy is based on both climate and energy policy considerations. Thus, reference is made to the double leverage effect that Europeanization as such produces, on the one hand, a clear will to Europeanize the national energy mix and, on the other hand, the relatively partial effect of this effort. Basically, the conceptual dimension of the Europeanization phenomenon allows us to better understand its scope. In this respect, the content of the general objective allows us to understand, for example, that it includes sectoral objectives and automatically excludes national objectives. Chapter 1 devotes a significant amount of time to the binding objective set by Directive (EU) 2018/2001, which establishes a collective obligation of result before encouraging a collective effort to achieve it through specific means. The central role of Directive (EU) 2018/2001 is regularly recalled insofar as it participates in the Europeanization of the promotion of the use of renewable energy (RE), and certainly contributes to the renewal of the European ambition to achieve the objectives of renewable energy consumption. The challenge is indeed to see how this truly collective commitment responds to the major challenges of energy law and more globally of the energy transition.

Chapter 2 provides an in-depth analysis of the European Union's (EU) external relations in the energy sector, through the interesting lens of EU-Russia trade relations and the Energy Community. Based on the assumption that securing reliable flows of various renewable energy sources would enhance the EU's energy security and reduce its vulnerability to major fossil fuel exporters, the developments enrich the doctrinal reflections on security of supply. In this sense, the reader is given a clear overview of the relevant data on EU energy production and consumption, coupled with the current trends and challenges facing the EU in the energy field. Chapter 2 also presents developments on EU treaty competence issues, and then examines the energy-related treaty regimes to which the EU belongs, while highlighting the EU's participation in the Energy Community ("EnC") and various related issues. Energy relations between the EU and Russia are analyzed from the perspective of oil and gas trade, thus providing ample information on current issues. The conclusions and recommendations reached in this chapter are remarkable. It shows that the EU's energy structure is highly dependent on external supplies. This fact explains why external energy relations play a crucial role in the development of foreign policy. "The ability to integrate an effective energy *acquis communautaire* in neighboring countries and to ensure a stable relationship with the largest energy supplier are two of the Union's main external objectives. However, to do this effectively while ensuring energy security requires certain policy solutions, which the EU has been implementing for some time: diversification of energy supplies, for example through liberalization of trade in green goods and services, promotion of renewable energy, and lobbying for modernization of the ECT."

Introduction

Chapter 3 focuses on the European Union's interest in offshore activities. It recalls that energy is at the heart of the historical dimension of European construction. That this interest, which is still so obvious, is a priori part of a sustainable development approach, closely linked to the integrated maritime policy, the blue economy and blue growth. Following the Deepwater Horizon disaster of April 20, 2010, the development of offshore oil and gas operations in marine waters under the sovereignty and jurisdiction of the Member States of the European Union, but also in neighboring waters, has imposed the subsequent need for better regulation and supervision. Therefore, the European Union's strategy implies being defined and implemented according to a double logic: it integrates internal and external action, in order to develop a dedicated legislative framework, as complete and efficient as possible, but also to pursue its international cooperation efforts in favor of a legal regulation of offshore activities, if not effectively global, at least beyond its borders, and in the regions and sea basins it shares with non-member States. The result is a new EU strategy for the legal governance of offshore energy activities, developed from Cooperation in Public International Law to Integration into European Union Law. With regards to cooperation in public international law, an in-depth analysis is made first from the material and/or formal framework of soft law, before being complemented by hard law. In the first case, the developments allow for a better understanding of the regional systems to which the EU is not a party, through the examples of the Arctic and the Black Sea, and then that of the regional treaties to which the EU is a party, such as the Baltic Sea or the OSPAR zone. Concerning the hard law, the analysis is becoming more and more detailed based on a series of instruments such as the protocol dedicated to offshore activities within the framework of the Barcelona system or the protocol on offshore activities of 1994. Thus, reference is made to the change of strategy and its decision to join the Offshore Protocol. Chapter 3 also returns to the integration in the European Union law, evoking the cross-foundation between the acquis of the European Union law and the Mediterranean acquis, including the implications that stem for the EU member states. In addition, a particular look is taken at the contribution of Directive 2013/30/EU on the safety of offshore oil and gas operations, before questioning the challenges of the EU's blue growth, whose limited normative contribution to legal enforceability remains relative.

Chapter 4 explores the important role that the state is still expected to play in initiating and implementing the energy transition. In this regard, it is presented in three dimensions. The first dimension focuses on the study of classical public law, particularly through the premise of the role of the state that derives from constitutional and administrative law. This role is considered classical because it is based on the different functions of the state, and the legitimate constraint that distinguishes it from other social actors, including non-state actors. Analysis from the perspective of sociology or legal studies, when dealing with the specific situation of French-speaking African states, offers considerable research material. The scope of the analysis is broadened by the approach to energy law, which allows us to contextualize and identify elements specific to energy law in a space where it could be in competition with classical public law. Focusing on English-speaking African countries, Chapter 4 examines both how the state adopts and enforces laws to design its own energy transition scheme and how it exercises discretion through its energy policy. Beyond the functions of the state-which derive from its sovereign power-these elements point the way toward a specific role the state can play in the energy transition as a process. As such, the energy transition, if it is to lead to coherent social change, requires strong and dynamic leadership, including a clear, nuanced, and forward-looking focus on the broad sections of the overall process, and the environmental justice issues that are necessarily associated with it. For this reason, the role of the state is interpreted as both leadership and integration of environmental, economic, and social issues. However, the necessary and strong support of international cooperation is also recalled here. The state—in order to achieve a smooth paradigm shift—needs to take advantage of the interactions between its internal efforts and the support it can draw from international cooperation. As a profound analysis is attempted in this chapter regarding the African state, emphasis is placed on the African Union's transition initiatives, which aim to have a dual effect: that of an incentive, on the one hand, and support, on the other hand, for state efforts. Taken together, these elements allow us to think of a coherent model of the role that African states play or could play in the energy transition.

More than a reflection on energy law and policy, this book offers the reader an interesting journey into the major issues of the discipline and the richness that can result from the crossing of views of scholars from different continents.

1

THE EUROPEANISATION OF RENEWABLE ENERGY CONSUMPTION TARGETS RENEWED BY DIRECTIVE (EU) 2018/2001

Bernadette Le Baut-Ferrarese

1.1 Introduction

If to Europeanize is all about theorizing a notion, a concept, an activity, etc. Whatever it might be, from the European Union ("EU") perspective, it is also about considering such thing as a notion, a concept, an activity, as susceptible to be enshrined by EU secondary law. Not necessarily by its primary law.[1]

With due regard to the principle of conferral, as set out in Article 5 of the Treaty on European Union ("TEU"), Europeanisation is no less contingent, regardless of its purpose, on the possibilities offered by European Union ("EU") treaties. In doing so, it must be noted that, especially when it comes to renewable energies ("RE"), the Treaty on the Functioning of the European Union ("TFEU") makes it legally possible, by allowing or explicitly considering it. For quite some time, the phenomenon of Europeanization—as regards REs—was thus carried out on legal bases—not necessarily diverted from their original purpose—but instrumentalized by the EU's institutions at the very least.

More accurately, on the premise that REs are beneficial to the environment, given their inexhaustible nature and environmental credentials in relation to greenhouse gas ("GHG") emissions, the EU chose to base the Europeanisation of these energies on the EU policy in the "area" of the environment (Article 191 et seq. TFEU; ex Article 174 et seq. TEC). This has led to the adoption of many emblematic instruments. Such is indeed the case with

---

[1] The apprehension of renewable energies by primary law consists in the possibility of applying to acts and/or activities that are related to said energies the devices of obligations, to do or not to do, that it contains. This phenomenon can be illustrated by the famous Vent de Colère case in which the CJEU (see 19 December 2013, Case C-262/12) and the French Conseil d'Etat (see 28 May 2014, Association Vent de Colère!, no. 324852; 15 April 2016, Association Vent de colère!, no. 324852), applied the Treaty provisions on the prohibition of State aid (TFEU, art. 107 et seq.) to a scheme put in place by France to support onshore wind power generation.

> Directive 2001/77/EC of the European Parliament and of the Council of 27 September 2001 on the promotion of electricity produced from renewable energy sources in the internal electricity market.[2]
>
> Directive 2003/30/EC of the European Parliament and of the Council of 8 May 2003 on the promotion of the use of biofuels or other renewable fuels for transport.[3]
>
> Directive 2009/28/EC of the European Parliament and of the Council of 23 April 2009 on the promotion of the use of energy from renewable sources and amending and repealing Directives 2001/77/EC and 2003/30/EC.[4]

In fact, it is only since its revision by the Treaty of Lisbon that the TFEU provides for a Europeanisation that specifically concerns REs, which finds its place in the policy "in the field of energy" that it enshrines (TFEU, Article 4 and Article 194) in the context of which it is intended to be embodied in "measures" promoting "… the development of new and renewable forms of energy" (Article 194.1 (c) TFEU). Measures of which Directive (EU) 2018/2001 of the European Parliament and of the Council of 11 December 2018 on the promotion of the use of energy from renewable sources[5] is the most recent incarnation.

The choice of targeting Europeanisation towards the "promotion" of REs is instructive: it reveals the EU's favour for a category of energies that are advocated—if not plebiscited—for their environmental advantages, but also—in this perspective—reflects either the EU's adherence or the participation—in the process of energy transition of which REs are consubstantial.[6] This orientation of Europeanisation seems exclusive of any other consideration. This observation can be drawn from Article 194.2 TFEU

---

2   OJEU 2001, L. 283, 33.
3   OJEU 2003, L. 123, 42.
4   OJEU 2009, L. 140, 16.
5   OJEU 2018, L. 328, 82.
6   This process is defined as 'the transition, in the long term, from non-renewable energies (oil, coal, gas, uranium) to renewable energies that can be accompanied by a reduction in greenhouse gas emissions, a reduction in environmental risks and a reduction in pollution and hazardous waste.' P. Sabliere, *Droit de l'énergie* (Dalloz, 2013) 46.

which—by prohibiting the EU from enacting measures whose purpose is to "determine the conditions for exploiting ... its energy resources" of the Member States—excludes from the scope of Europeanisation important parts of the law likely to be applied to REs, and—thereby—makes the advent of a broad-spectrum EU law for these energies illusory.

Therefore, dedicated to the promotion of the development of REs, the Europeanisation of the former appears to be a process. Indeed, it takes place progressively, by means of texts forming a legal continuum. The duration of their application is indeed limited to the current or ensuing decade after their adoption, while their extent[7] and/or scope[8] are evolving. This process is rather flexible, since it is based on directives. In other words, on texts that are supposed to provide Member States with a margin of discretion as to their transposition (Article 288 TFEU). In short, this process suggests a Europeanisation of national promotion rather than a European promotion *stricto sensu* of the development of REs.

Directive (EU) 2018/2001/EU, together with its corrigendum,[9] is the latest version of the above-mentioned Europeanisation process. Adopted within the framework of the legislative package "Clean energy for all Europeans," the directive comes chronologically after directive 2009/28/EC, which it "repeals" (Article 37), and of which it ensures the "recasting" for the sake of "clarity."[10] Directive (EU) 2018/2001/EU also takes up the object of directive 2009/28/EC, aiming at establishing a "common framework for the promotion of energy from renewable sources" (Article 1). In particular, the new directive develops the common framework

---

7   See in this sense, Dir. 2001/77/EC, Art. 1: 'the purpose of this Directive is to encourage an increase in the contribution of renewable energy sources to the production of electricity in the internal electricity market and to lay the foundations for a future Community framework in this area.'

8   See in this sense the titles of the various directives concerned here: Dir. 2001/77/EC, which refers to 'electricity from renewable sources,' then Dir. 2003/30/EC, which refers to 'biofuels and other renewable fuels used in transport' (see Dir. 2003/30/EC cited above), then Dir. 2009/28/EC cited above and Dir. 2018/2001 cited above, which refer to 'energy from renewable sources' i.e., *a priori*, they now refer to all forms or types of energy that make use of renewable energy sources.

9   See Corrigendum to Directive (EU) 2018/2001 of 11 December 2018 on the promotion of the use of energy from renewable sources: OJEU 2019 L.75, 137.

10 This is due to the numerous amendments made to Directive 2009/28/EC after its adoption: see in this sense Dir. (EU) 2018/2001, recital 1.

relating to the objectives of RE consumption, which is far from exhausting the subject of Europeanization,[11] but which has always constituted an essential part of it until now.[12] More precisely, on the one hand, it sets "a binding Union target for the overall share of energy from renewable sources in the Union's gross final energy consumption in 2030" (Article 1); on the other hand, it specifies that "Member States shall collectively ensure that the share of energy from renewable sources in the Union's gross final energy consumption in 2030 is at least 32%" (Article 3.1). In other words, Directive (EU) 2018/2001 renews the Europeanisation by modifying both the level and the scope of the said targets.

1.2 The Level

Directive (EU) 2018/2001 seeks to raise the EU's level of ambition with regards to renewable energy consumption targets. It continues the process of Europeanisation in this respect. On the one hand, it confirms the goal of increasing the share of REs in the overall final energy consumption in the EU, whereas, on the other hand, it strengthens the scope of increasing the share of these energies in the national energy mix.

---

11 As stated in Article 1, Directive (EU) 2018/2001 '[lays down] rules on financial support for electricity from renewable sources, on self-consumption of such electricity, on the use of energy from renewable sources in the heating and cooling sector and in the transport sector, on regional cooperation between Member States, and between Member States and third countries, on guarantees of origin, on administrative procedures and on information and training. It also establishes sustainability and greenhouse gas emissions saving criteria for biofuels, bioliquids and biomass fuels.' For a more comprehensive analysis of this text, see our article, 'La directive 2018/2001 du 11 décembre 2018 relative à la promotion de l'utilisation des énergies renouvelables: entre renouvellement de l'européanisation et européanisation renouvelée' (2019) *Énergie - Env. - Infrastr.* Dossier 27.

12 In particular, it appears to be the decisive element enabling the EU to justify the extent to which its competence on the subject can be exercised in the light of the principle of subsidiarity; see, in this sense, Recital 128 of Directive (EU) 2018/200: 'Since the objective of this Directive, namely to achieve a share of at least 32 % of energy from renewable sources in the Union's gross final consumption of energy by 2030, cannot be sufficiently achieved by the Member States but can rather, by reason of the scale of the action, be better achieved at Union level, the Union may adopt measures, in accordance with the principle of subsidiarity as set out in Article 5 of the Treaty on European Union. In accordance with the principle of proportionality, as set out in that Article, this Directive does not go beyond what is necessary in order to achieve that objective.'

1.2.1 Increase in the Share of Renewable Energy in Overall Final Energy Consumption

Article 3.1 of the new directive sets out the objective of increasing "the share of energy from renewable sources in the Union's gross final consumption of energy in 2030" to "at least 32%" (Article 3.1). This is a policy choice, more precisely an energy policy choice, determined by climate policy issues.

1.2.1.1 Energy Policy Choices

The inclusion of quantitative targets for the consumption of REs in EU legal instruments promoting the use of REs is a choice of public policy, and—in this case—of energy policy. More precisely, it is a choice that implements a vision that is—to say the least—comprehensive of the idea of "promotion," which is at the heart of all the directives that have so far enshrined it, and which—according to the terms of Article 194.1 (c) TFEU—characterises the current competence of the EU on the subject, even though, in the framework of the EU, promotion actions are theoretically not

> anything other than actions of encouragement, support, and assistance. They are traditionally chosen and attributed to maintain a European framework of accompaniment to competences reserved to the States.[13]

The choice reiterated by Directive (EU) 2018/2001 stems—in this case—from the political and institutional agreement reached under the "ordinary legislative" procedure, which Article 194 TFEU makes applicable here. While this increase in the target for RE consumption in the EU was, ab initio, the result of a legislative proposal by the Commission, which plays a decisive role in guiding the EU's major energy choices,[14] the choice of setting it at 32% for 2030 is, above all, the compromise reached in the context of the co-legislative work carried out by the European Parliament and the Council, even though the Parliament had initially expressed the wish to set it at a level of 35%.[15]

---

13 F. Peraldi-Leneuf, 'La politique européenne sur les ENR et l'efficacité énergétique : éclatement des responsabilités ou politique intelligente' in C. Bluman (ed.), *Vers une politique européenne de l'énergie* (Bruylant, 2012) 88.

14 TUE, Art. 17. : 'La Commission promeut l'intérêt général de l'UE et prend les initiatives appropriées à cette fin.'

15 C. Lepage, 'Les différents accords sur l'énergie intervenus au niveau européen

1.2.1.2 Climate Policy Choices

The target of 32% of RE consumption by 2030 set by Directive (EU) 2018/2001 is also a political choice because it is based on climate policy considerations. More specifically, it is the result of the EU taking into account the constraints on the subject arising from international law.

In 2002, the EU ratified the Kyoto Protocol to the United Nations Framework Convention on Climate Change ("UNFCCC").[16] As a result, the EU has taken various measures including on a GHG emission allowance trading scheme, on effort-sharing between Member States, on carbon capture and storage, on the quality of petrol and diesel fuels and on the reduction of CO2 emissions from motor vehicles, and on accounting rules for GHG emissions and removals from land use, land-use change and forestry activities. Then, in 2016, the Union ratified the Paris Agreement[17] as adopted by the Conference of the Parties to the UNFCCC in December 2015, which pursues the objective of containing the rise in global temperature to between 1.5° C and 2° C above pre-industrial levels and enshrines the idea of climate neutrality,[18] and has the particularity of being centred on the concept of "nationally determined contributions."[19]

---

sont-ils suffisamment ambitieux?' (2018) Actu-environnement.com <https://www.actu-environnement.com/blogs/corinne-lepage/77/corinne-lepage-accord-energie-europe-ambition-112.html> (accessed 26 June 2018).

16   See 2002/358/EC: Council Decision of 25 April 2002 concerning the approval, on behalf of the European Community, of the Kyoto Protocol to the United Nations Framework Convention on Climate Change and the joint fulfilment of commitments thereunder, OJ L 130, 15/05/2002, 0001-0003.

17   See Council Decision (EU) 2016/1841 of 5 October 2016 on the conclusion, on behalf of the European Union, of the Paris Agreement adopted under the United Nations Framework Convention on Climate Change OJ L 282, 19/10/2016, 1–3.

18   Article 4.1 states that 'Parties aim to reach global peaking of greenhouse gas emissions as soon as possible, recognizing that peaking will take longer for developing country Parties, and to undertake rapid reductions thereafter in accordance with best available science, so as to achieve a balance between anthropogenic emissions by sources and removals by sinks of greenhouse gases in the second half of this century ...'

19   See Article 4.2, which provides that 'Each Party shall prepare, communicate and maintain successive nationally determined contributions that it intends to achieve. Parties shall pursue domestic mitigation measures, with the aim of achieving the objectives of such contributions.'

The EU has not only signed and ratified these international agreements ("IA"), it has also adapted its law to the clauses contained in those IA. Although it was not a pioneer in this field, the Community and then the EU have thus established themselves as a central player, if not the leader, in the fight against climate change, a move that is all the more remarkable given that primary law does not enshrine any European climate policy. In fact, as is often the case in the EU framework, this adaptation is the result of a dynamic reading of the competences attributed, based on the so-called "integration of environmental protection into the definition and implementation of the Union's policies" clause (Article 11 TFEU), based on the observation that climate change is above all an environmental problem.

While the EU is striving to insert a climate dimension into all its policies, because the production and use of fossil fuels are the main sources of GHG emissions that cause climate change, the EU has chosen in particular to integrate the climate issue into the actions it is required to take in the energy sector. In fact, the EU's energy policy is above all an "integrated" energy-climate policy, as it is overlaid by European objectives for the internal reduction of GHG emissions. Precisely, the legislative packages it adopts in the field of energy are built around the ambition of reducing overall internal GHG emissions: Thus, the 2009 legislative package known as "Energy-Climate" aimed to reduce GHG emissions for 2020 by at least 20%;[20] then, the 2018-2019 legislative package "Clean Energy for all Europeans" refers to the objective of a reduction of at least 40% of the GHG emissions by 2030, in this case, as a specific "contribution" of the Union and its Member States to the Paris Agreement.[21]

It is because RE have an essential role to play in the success of these commitments that the EU can link its directives promoting them to these ambitions.

Recital 1 of Directive 2009/28/EC thus stated that

> the control of European energy consumption and the increase in the use of energy from renewable sources con-

---

20 See in this sense Recital 3 of Directive 2009/29/EC of the European Parliament and of the Council of 23 April 2009 amending Directive 2003/87/EC, so as to improve and extend the greenhouse gas emission allowance trading scheme of the Community (Text with EEA relevance) OJ L 140, 5.6.2009, 63–87.

21 European Council: Conclusions of 23-24 October 2014 on the 2030 climate and energy policy framework; Conclusions of 17-18 March 2016.

stitute, together with energy saving and increased energy efficiency, important elements of the package of measures required to reduce greenhouse gas emissions and to comply with the Kyoto Protocol to the United Nations Framework Convention on Climate Change, as well as with other commitments made at Community and international level with a view to reducing greenhouse gas emissions beyond 2012.

The Directive (EU) 2018/2001 now states that

> increasing the use of energy from renewable sources (...) is an important element of the energy package.) is an important element of the package of measures required to reduce greenhouse gas emissions and to comply with the Union's commitments under the 2015 Paris Agreement on climate change.[22]

If there is a direct link between the EU's energy objectives, in particular those relating to the consumption of RE, and its climate objectives,[23] it is safe to understand why the development of the latter necessarily affects that of the former. This is the meaning of the review clause provided for in Article 3.1 of Directive (EU) 2018/2001, which allows the Commission to update the target of 32% RE consumption in the EU by 2030, by presenting

---

22 In the same sense, see also Recital 1 of Directive 2001/77/EC: 'The potential for the exploitation of renewable energy sources is underused in the Community at present. The Community recognises the need to promote renewable energy sources as a priority measure given that their exploitation contributes to environmental protection and sustainable development. In addition this can also create local employment, have a positive impact on social cohesion, contribute to security of supply and make it possible to meet Kyoto targets more quickly ...' or Art. 1 of the above-mentioned Directive 2003/30/EC: 'This Directive aims at promoting the use of biofuels or other renewable fuels to replace diesel or petrol for transport purposes in each Member State, with a view to contributing to objectives such as meeting climate change commitments, environmentally friendly security of supply and promoting renewable energy sources.'

23 'This 32% is close to the 33% that, according to a recent study, would allow, combined with 33% energy efficiency, to reduce the Union's greenhouse gas emissions by 46% by 2030.' See C. Lepage, 'Les différents accords sur l'énergie intervenus au niveau européen sont-ils suffisamment ambitieux?' (n 15).

by 2023 a legislative proposal to increase it in the event of a further significant fall in the costs of renewable energy production, if this is necessary in order to comply with the international commitments made by the Union with regard to decarbonization.[24]

A clause that the EU could be led to activate[25] in the event that the European Parliament and the Council endorse its proposal for an upward revision of the European GHG emissions reduction target for 2030.[26]

In any case, the existence of such a link raises questions about the adequacy of the former and the latter and, consequently, about the possibility of challenging the choice made by the EU before the courts, in this sense in the context of litigation before the Court of Justice of the European Union ("CJEU"), which—by definition—will have a climate dimension. As it stands, this hypothesis raises two types of difficulties. The first is procedural. It relates to the possibility of access to the Court of Justice of the European Union (CJEU) for individuals who, challenging the RE consumption objective enshrined in an EU directive, would do so in the context of the annulment proceedings provided for in Article 263 TEU, even though these texts are de facto virtually inaccessible to such proceedings.[27] The Court of First Instance of the EU has moreover

---

24 In the same sense, but with regards to the specific target for renewable energy consumption in transport, see also EU Directive 2018/2001, Art. 25.1.

25 'D'ici 2030, les engagements des États membres devraient permettre d'atteindre 33,7 % d'énergies renouvelables, soit légèrement plus que l'objectif fixé à 32 % au minimum. The new climate target should reach 38-40%.' S. Fabregat, 'Climat : le plan de la Commission pour réduire les émissions de carbone de 55 % d'ici 2030,' (2020) Actu-environnement.com <https://www.actu-environnement.com/ae/news/Climat-union-europeenne-reduction-ges-transport-batiment-36122.php4> (accessed 17 September 2020).

26 See U. Van Der Leyen, 'State of the Union Address' (EC Press corner, 16 September 2020) <https://ec.europa.eu/commission/presscorner/detail/en/SPEECH_20_1655> (accessed 16 September 2020).

27 Article 263 TFEU requires that so-called ordinary applicants, i.e., private individuals or persons, who act against an act of general application be 'individually' affected by it. The CJEU gave a restrictive interpretation of this condition in its *Plaumann* judgment of 15 July 1963: 'persons other than the addressees of a decision may claim to be individually concerned only if that decision affects them by reason of certain qualities peculiar to them or of a factual situation which distinguishes them from any other person and thereby individualises them in a manner analogous to that of the addressee.' See Case C-25/62, ECR 1963, I, 197.

ruled in this sense, having declared inadmissible the action for annulment brought against the provisions of Directive (EU) 2018/2001 by individuals from various Member States of the European Union and the United States and associations with their headquarters in various Member States who argued before it that the provisions of this text were incompatible with the climate objectives of the EU,[28] on the grounds that they had not been able to demonstrate to what extent it affected them "individually."[29] The second difficulty relates to the political dimension of the objectives in question, which invites one to admit that the EU institutions in charge of normative power have—on the subject—a broad power of appreciation, and consequently to refuse to allow the EU court any possibility of calling them into question, supposing that it could be validly seized of the matter,[30] at the risk of otherwise engaging in a control of opportunity.

1.2.2 Increase in the Share of Renewable Energy in the National Energy Mix

It is also clear that, from the moment that the promotion of the development of REs includes objectives of consumption of these energies, it comes to Europeanize not only a part but also partially the national energy mix.[31]

---

28 In the present case, the applicants argued that the inclusion of forest biomass in the list of renewable energy sources would have been detrimental to the attainment of the objectives of Directive (EU) 2018/2001, because of the amount of carbon released by the combustion of wood and the increase in industrial forestry.

29 See Trib. EU, Order, 6 May 2020, Case T. 141/19, Peter Sabo and Others v European Parliament and Council.

30 This could be the case in the following two hypotheses: that of an action for annulment brought by the Member States and/or by the EU institutions, which are in a privileged position compared to ordinary claimants, since they always have the right to bring such an action against any challengeable EU Act (TFEU, Art. 263(2)); that of a preliminary ruling on the validity of the directive by a national court (TFEU, Art. 267 TFEU). The Court of First Instance of the European Union was careful to mention this possibility in paragraph 43 of its order Peter Sabo and Others v. European Parliament and Council of 6 May 2020.

31 The energy mix is defined as the distribution of the different primary energy sources consumed in a given geographical area (source: Wikipedia France). In other words, it covers 'all the energies, and their best possible proportions, needed to cover a country's needs, taking into account consumption in the transport, industry, trade and agriculture sectors, as well as in the public and household sectors.' P. Sabliere, *Droit de l'énergie*, (n 6) 3.

1.2.2.1 Europeanisation of Part of the National Energy Mix

It is a fact that the EU's approach to promoting the use of REs has included quantitative targets for the consumption of these energies in the EU since its inception. These targets were first set for "electricity from renewable sources" (Directive 2001/77/EC), then for "biofuels and other renewable fuels for transport" (Directive 2003/30/EC), and now for "energy from renewable sources" (Directive 2009/28/EC and Directive (EU) 2018/2001).

Admittedly, the scope of the energies concerned by said objectives has always been *de facto* limited, more precisely limited to those that call upon renewable sources as listed in the directives.[32] It is no less legitimate to question the Europeanisation process when, as is the case in this instance, it results in nothing more or less than imposing on Member States the introduction of RE sources into their energy mix. The observation is indeed implacable: what was prima facie announced as a "simple promotion of renewable energies,"[33] leads de facto to "force(d) all Member States to modify their energy mix."[34]

This observation is all the more remarkable when it is made with regard to Directive (EU) 2018/2001, which—as the most recent milestone in the Europeanisation of the promotion of the development of renewable energies—is also the first major EU text in this process to be based on Article 194 TFEU, i.e., on a provision whose paragraph 2 explicitly reserves to the State the "choice between different energy sources." In other words, if this directive deserves attention, it is because by Europeanizing RE consumption targets, it not only comprehensively applies paragraph 1 of this provision, but also flexibly applies paragraph 2.[35] It is indeed dif-

---

32   See most recently dir. (EU) 2018/2001, Art. 2.1: 'energy from 'renewable sources' or 'renewable energy' means energy from renewable non-fossil sources, namely wind, solar (solar thermal and solar photovoltaic) and geothermal energy, ambient energy, tide, wave and other ocean energy, hydropower, biomass, landfill gas, sewage treatment plant gas, and biogas.'

33   F. Berrod and A. Ullestad, *La mutation des frontières dans l'espace européen de l'énergie* (Larcier, coll. Paradigme, 2016) 125.

34   Ibid.

35   On this topic, see also A. Johnston and E. Van Der Marel, 'Ad lucem? Interpreting the New Energy Provision, and in Particular the Meaning of 194(2) TFEU' (2013) 22 *European Energy and Environmental Law Review* 181.

ficult not to consider the energy sovereignty of the State as undermined when de facto the one remaining to the State as regards the choice of the energy sources exploited on its territory is limited to decide, among those which have a renewable origin, those whose exploitation seems to him the most appropriate in order in particular to satisfy the European objective of consumption envisaged by the directives of the EU?[36] In short, these texts convey an extensive vision of the European approach to energy integration which, although permitted by the TFEU, does not seem to be permitted by the EURATOM Treaty. Indeed, this treaty—which de jure does not "promote" nuclear energy—[37] but which nevertheless de facto entrusts the European institutions "with the task of promoting research into and the peaceful use of nuclear energy ... [but] does not, however, imply that the European Union should be a party to the Treaty on European Union.[38] However, this does not imply that the Member States use nuclear energy, nor does it authorise the Community legislator to prescribe the use of nuclear energy.[39]

1.2.2.2 Partial Europeanisation of the National Energy Mix

It must be acknowledged that if the approach consisting in setting EU-wide RE consumption targets Europeanizes part of the national energy mix, it also manages to partially Europeanize the latter, for the simple reason that, by implication, it implies the reduction within it of the share of fossil and/or nuclear energy.

---

36 See in this sense ECJ, 10 Sept. 2009, Plantanol GmbH, Case C-201/08, pt 37 (concerning the above-mentioned Directive 2003/30/EC): '... the Member States also enjoy a wide discretion with regard to the products which they wish to promote in order to attain the objectives laid down in the directive, since they may choose to give priority to the promotion of certain types of fuels by taking account of their overall cost-effective climate and environmental balance, while also taking into account competitiveness and security of supply.'

37 It states in its preamble that it aims to create the conditions for the development of a strong nuclear industry, and provides in its Article 1, second paragraph, that 'the task of the Community to contribute to the raising of the raising of the standard of living in the Member States and to the development of relations with other countries by creating conditions necessary for the speedy establishment and growth of nuclear industries.'

38 G. Winter, 'L'ascension et la chute de l'utilisation de l'énergie nucléaire en Allemagne: les processus, les explications et le rôle du droit' (2014) 39 *Revue juridique de l'environnement* 231, 238.

39 Ibid.

If this approach deserves to be questioned, it is a fortiori when it is implemented by a text such as Directive (EU) 2018/2001, i.e., based on Article 194 TFEU, paragraph 2 of which reserves for the State the competence to determine "the general structure of energy supply."[40] In fact, this provision, which aims to protect the energy sovereignty of the State, appears to limit the scope of the Europeanisation of the national energy mix. More precisely, it raises the question of the maximum proportions of RE consumption targets when they are established by EU law. The fact is that the TFEU does not allow to know the maximum intensity of the Europeanisation of RE consumption targets, which one can, however, expect to respect reasonable limits, if only to be politically accepted by the Member States.[41] As a result, there is some uncertainty as to the future development of this aspect of the Europeanisation of the promotion of the use of RE, even though this approach appears to be a process—in other words—a promise of successive advances.

1.3 The Scope

The scope of the RE consumption targets has varied as the promotion of the development of REs has become more European. For its part, Directive (EU) 2018/2001 states—in its Article 3—that a "global" and "binding" target should be established.

1.3.1 Overall Objective

The "global" target for RE consumption enshrined in Directive (EU) 2018/2001 is twofold: on the one hand, it includes a sectoral target, and on the other hand, it does not include national targets.

1.3.1.1 Including a Sectoral Target

The sectoral dimension of the Europeanisation of RE consumption tar-

---

40 See also in the same sense, European Council 12-13 Dec. 2019, Conclusions: 'EU leaders acknowledged the need to ensure energy security and to respect the right of the member states to decide on their energy mix and to choose the most appropriate technologies. Some countries have indicated that they use nuclear energy as part of their national energy mix.'

41 While it seems clear that Europeanization cannot lead to an obligation on Member States to establish an energy mix based on 100% renewable energy sources, the level of shares of these energies that it can achieve without risking infringement of Article 194.2 TFEU remains unknown and therefore uncertain.

gets has always been present. Ab initio, it will take place in directives promoting REs, which are themselves sectoral in scope: so it is with directive 2001/30/EC on the promotion of the use of renewable electricity in the internal market,[42] and with directive 2003/30 aimed at promoting the use of biofuels or other renewable fuels in transport.[43]

The 2009/28/EC directive puts an end to the sectoral Europeanisation of the promotion of Res, but does not abandon a sectoral approach when it comes to the Europeanisation of the consumption objectives of these energies. In fact, this directive, aiming to promote these energies in general, includes a European target for renewable energy consumption specific to the transport sector, stating that "each Member State shall ensure that the share of energy from renewable sources in all forms of transport in 2020 is at least equal to 10% of its final energy consumption in the transport sector" (Article 3.4). Directive (EU) 2018/2001, which is in its wake, reiterates the same rationale, stating that "in order to integrate the use of Re in the transport sector, each Member State shall impose an obligation on fuel suppliers to ensure that by 2030 the share of renewable energy in final energy consumption in the transport sector reaches at least 14% (minimum share), in accordance with an indicative trajectory defined by the Member State in question" (Article 25.1).

These last two versions of the Europeanisation of RE consumption targets can nevertheless be considered as a step backwards on the matter. Indeed, on the one hand, they no longer include a Europeanized renewable electricity consumption target. On the other hand, they no longer seem to see the sectoral Europeanisation approach applied to other segments of RE production. In this sense, even though Directive (EU) 2018/2001 takes care to emphasize the importance of the heat and cooling sectors, in terms of weight in the consumption of RE and final energy in the EU, recalls their interest in the "decarbonization" and "security" of energy in the EU, and finally goes so far as to regret "the relatively slow progress" of the Member States on the subject.[44] It does not, however, go so far as to

---

42 This Directive stated an indicative share of 22.1% of electricity produced from renewable energy sources in total Community electricity consumption in 2010: see Dir. 2001/77/EC, Art. 3.4.

43 This Directive set a renewable fuel consumption target of 5.75% for 2010: see Dir. 2003/30/EC, Art. 3.

44 See EU Directive 2018/2001/EU, para. 73.

establish European consumption objectives[45] for these sectors, but—on the contrary—noting the "absence of a harmonised strategy at the level of the Union," it limits itself to encouraging the Member States to develop these sectors and to set the level of effort to be made at their level.[46] In this sense too, the same directive does not propose any specific consumption objectives for the renewable gas[47] or hydrogen sectors.

1.3.1.2 Does Not Include National Targets

Although Directive (EU) 2018/2001 is a text that participates in the process of Europeanisation of the promotion of the use of REs, it differs from it in that it does not specify the "global" objective of consumption of the said energies that it enshrines for each Member State.

However, the principle of a quantified distribution of the "global" effort to be made, Member State by Member State, seemed to be inherent to the Europeanisation process until now. In fact, it had been applied: first, by Directive 2001/77/EC, which had established "national indicative targets" (Article 3.1) for renewable electricity consumption,[48] accompanied by "reference values;" (Annex) then, by Directive 2009/28/EC, which had adopted "overall national binding targets" (Article 3.1) for RE consumption, presented in a "table" (Annex I, Part A) and explained by an indicative trajectory (Annex I, Part B).

By refusing to repeat this approach, the framers of Directive (EU) 2018/2001 have made a choice which, although legally possible—the EU's competence in the field of energy being shared, and therefore reversible—[49] nonetheless raises questions about its motives. Firstly, it is legitimate to think that this choice may have resulted from the Member

---

45   However, with regards to heating and cooling, the provisional agreement on the directive provided for a sub-target of an indicative increase of 1.3% per year in renewable energies calculated over the 5-year period starting in 2021; see C. Lepage, 'Les différents accords sur l'énergie intervenus au niveau européen sont-ils suffisamment ambitieux?' (n 15).

46   See EU Directive 2018/2001/EU, Cons. 74 and Art. 23.1.

47   However, these can be taken into account when calculating the overall share of gross final consumption of energy from renewable sources: see in this sense Dir. (EU) 2018/2001, Art. 7.1.

48   See Directive 2001/77/EC, Article 3.1.

49   In the case of shared competences, the EU is in fact free to go back, i.e., to relinquish all or part of a pre-empted competence.

States' desire to (re)nationalise the RE consumption targets,[50] whether they were anxious to ensure compliance with the letter of a treaty which now explicitly recognises their competence to determine the content of their national energy mix (TEU, Art. 194.2), or whether they wished to protect themselves against the "arbitrariness" of the method used by the EU to set the national renewable energy consumption targets.[51] Secondly, one must bear in mind the results obtained by the approach of individualising the overall target for renewable energy consumption in terms of efficiency, i.e., the actual achievement of the Europeanised national target by each Member State,[52] but also—by extension—the sanctioning of the Member State if it has not achieved it, are the result of a retrospective criticism.

If the implementation of a sanction against the State failing to achieve its national target for RE consumption is possible, it is—however—only if the legislation that carries it qualifies it as "binding," which was precisely the case with Directive 2009/28/EC. The simple finding of the existence of a "violation by the State of its obligations under the directive,"[53] thus

---

50 From this point of view, Directive (EU) 2018/2001 follows the pattern of Directive 2003/30/EC, which stated that 'Member States should ensure that a minimum proportion of biofuels and other renewable fuels is placed on their markets ...' and which provided that they '... shall set national indicative targets' (Art. 3.1, a.).

51 For example, the binding national targets established by Directive 2009/28/EC were set as follows: all Member States were required to increase their target for renewable energy consumption in final energy consumption by 5.7%, with the remainder being modulated according to the gross domestic product of each Member State. According to this process, for example in France, renewable energies were to represent 23% of total energy consumption in 2020.

52 According to Eurostat statistics for 2019: renewables covered 19.7% of the gross final energy consumption of the 27 EU member states, with a target of 20% renewables in 2020. France reached 17.2% of renewables. It is 5.8 points away from its national target of 23% in 2020. France is the country furthest behind, along with the Netherlands (5.2 points behind), Ireland, and Luxembourg (4 points behind their target). In contrast, 14 Member States had exceeded their national targets for 2020 and six countries were close to their targets.

53 See in particular: CJEU, 20 September 2017, Elecdey Carcelen, Case jtes C-215/16 and C-216/16, C-220/16 and C-221/16, para. 40: '... even assuming that it were accepted that that levy ... is capable of meaning that the Member State concerned does not comply with the mandatory national overall target set out in part A of Annex I to Directive 2009/28, the result would be ... an infringement, by that Member State, of its obligations under that directive...' (11 July

made it possible to conceive the principle of such a sanction, but without certainty as to its commitment or its outcome.

On the one hand, while in the framework of the EU legal order, such a sanction must be envisaged in the form of an action for failure to fulfil obligations (TFEU, Arts. 258 to 260), it must be remembered that the initiative for this procedure depends on the discretionary assessment of the Commission and/or a Member State, and that its outcome is only a judgment of the CJEU which, limiting itself to establishing the possible failure of a State to fulfil its obligations, can only do so ex post, i.e., necessarily (too?) late in the circumstances of the case.[54]

On the other hand, if it is necessary to recall the principle according to which the State engages its responsibility before the national judge when it violates the law of the EU,[55] it is necessary to admit that in this case it will be difficult to establish. Firstly, the conditions laid down by the case law of the CJEU to be able to engage this liability[56] seem difficult to meet here: that of a sufficiently serious breach of EU law, which does not seem to be serious here since the situation exposed here clearly corresponds to an incorrect transposition, and not to a non-transposition, of a directive;[57]

2019, Agrenergy, Case C-180/18), para. 27: '(...) Member States are not under any obligation, for the purposes of promoting the use of energy produced using renewable sources, to adopt support schemes. They therefore have discretion as to the measures they consider appropriate for the purpose of reaching the mandatory national overall targets set in Article 3(1) and (2) of the directive, read in conjunction with Annex I thereto... *Une telle marge d'appréciation implique que les États membres sont libres d'adopter, de modifier ou de supprimer des régimes d'aide, pourvu, notamment, que ces objectifs soient atteints*' (*souligné par l'auteur*); 28 mai 2020, Eco-Wind Construction, Case C-727/17, paras. 63 à 65.

54 The judgment will in fact occur at the end of the period of application of the directive, i.e. when new European legislation promoting the use of renewable energy sources will have been adopted, with its own, new, deadline for implementation.

55 In its judgment of 19 November 1991, Francovich and Bonifaci (Case 6/90), the CJEU ruled that the State must make good the harmful consequences of violations of Union law for which it is responsible, and that it is up to the national court to ensure that this is done by means of a liability action.

56 Conditions arising from its judgment of 5 March 1996 in Case C-46/93 *Brasserie du pêcheur v Bundesrepublik Deutschland* and *The Queen v Secretary of State for Transport, ex parte Factortame and Others*.

57 See in this sense ECJ, 8 October 1996, Case C-178/94, *Dillenkofer and Others v Bundesrepublik Deutschland*.

but also that requiring that the violated EU norm confers subjective rights to individuals, which prima facie do not seem to derive from the provision of Directive 2009/28/EC under consideration here. Secondly, the conditions laid down by State law, which is applicable to such an action for damages in the name of national[58] procedural autonomy, may also complicate the task of the claimants, for example, in the area of the causal link between the damage and the loss.

The proof of this is the recent judgment of the Administrative Court of Paris, which, after having noted the "failure" of the French State to achieve, in this case, its national objectives in terms of improving energy efficiency and increasing the share of REs in gross final energy consumption, which it took care to emphasise had "contributed to the failure to achieve the objective of reducing greenhouse gas emissions," refused—however—to consider that it had contributed to aggravating the ecological damage for which the applicant associations were seeking compensation.[59]

1.3.2 Binding Target

If Directive (EU) 2018/2001 differs from Directive 2009/28/EC, it is by the fact that it not only sets an exclusively "global" RE consumption target but also qualifies it as "binding."[60] In doing so, this directive sets the Europeanisation of the promotion of the development of the said energy on a new path, by (1) enshrining a collective obligation of result and (2) by Europeanizing some of the means of its realization.

1.3.2.1 Collective Obligation of Result

Article 3(1) of Directive (EU) 2018/2001 reads as follows: Member States "shall collectively ensure that the share of energy from renewable sources in the Union's gross final consumption of energy in 2030 is at least

---

58 This principle, which is based on the case law of the CJEU (see 16 Dec. 1976, *Rewe v Landwirtschaftskammer für das Saarland*), means that the procedural rules for legal actions based on EU law are laid down in national law.

59 TA Paris, *Association Oxfam France, Association Notre Affaire à tous, Association pour la natur' et l'homme, Association Greenpeace FranceFrance* (2020), Cases 1904967, 1904968, 1904972, 1904976/4-1.

60 This feature is all the more interesting as it does not apply, on the other hand, to the EU target for energy efficiency, which continues to remain a poor relation of EU energy policy, see in this sense Art. 1 Dir (EU) 2018/2002 of 11 December 2018 amending Directive 2012/27/EU on energy efficiency.

32%." This text therefore literally enshrines an obligation of a collective nature which, addressed to all Member States, also implies an individual commitment on the part of each of them, in two complementary ways.

On the one hand—and in any case—Directive (EU) 2018/2001 must, like any other EU directive, be transposed into national law.[61] This implies that each Member State must take the necessary measures or steps to ensure that its national law allows the "global" and "binding" European target RE consumption to be achieved. This seems to imply in fact that the objectives that each State may—at its own level—be led to set for the development of RE on its territory are compatible and consistent with it.[62]

On the other hand, the implementation of the collective obligation is intended to be developed in a more specific way within the framework of a "partnership" which "combines [the] actions [of the States] at the national level."[63]

In fact, Directive (EU) 2018/2001 presents a rather original legal situation, by enshrining an obligation of result of a collective nature, whereas EU directives are supposed to address to Member States only obligations of result of an individual scope.

1.3.2.2 Partial Europeanisation of Resources

The new directive on the promotion of the use of energy from renewable sources also stipulates, in Article 3(2), that it is up to the Member States to set "national contributions in order to collectively achieve the Union's overall binding target ... in the context of their integrated national energy and climate plans."

---

61 Directive (EU) 2018/2001/EU, Art. 36: (Member States are obliged to bring) 'into force the laws, regulations and administrative provisions necessary to comply with Art. 2 to 13, Articles 15 to 31, Article 37 and Annexes II, III and V to IX by 30 June 2021 at the latest.'

62 This is the case for national objectives [see, for example, C. énergie, art. L. 100-4 I.-4°: (France plans) 'To increase the share of renewable energies to 23% of gross final energy consumption in 2020 and to 32% of this consumption in 2030; by this date, to achieve this objective, renewable energies must represent 40% of electricity production, 38% of final heat consumption, 15% of final fuel consumption and 10% of gas consumption', but also, if they exist, regional objectives.

63 See Proposal for a Directive on the promotion of the use of energy from renewable sources (recast), COM (2016)767 final, Explanatory Memorandum.

The "integrated national energy-climate plan" ("INECP") appears to be the key instrument for the success of Europeanisation in that it constitutes an official formalization of the level of contribution of each State to the "global binding" objective established by Directive (EU) 2018/2001. This is undoubtedly the reason why the latter determines part of the content: by inviting each State to include "high ambitions" (Art. 3.5); by imposing European calculation rules for the elaboration of the national objectives that will be included (Art. 7); or by requiring that this document contain the "binding national targets" resulting from Directive 2009/28/EC, which it qualifies as "reference shares" of an incompressible and irreversible nature for the period 2021-2030 (Art. 3.4).[64] The content of Directive (EU) 2018/2001 appears to be rather unusual, insofar as EU directives do not normally impose obligations of means on Member States.

In addition, these provisions of Directive (EU) 2018/2001 take advantage of the European governance mechanism as established by Regulation 2018/1999 on the governance of the Energy and Climate Union.[65] This text appears to be particularly important in this context, and even innovative, insofar as it provides the Commission with means, both preventive and repressive, to guarantee the effectiveness of the Europeanisation of RE consumption targets, for example by requiring States to take additional measures if it considers that the global and binding target may not be met.[66] This "governance" regulation is also interesting in that it conceives the above—mentioned collective partnership in compliance

---

64　See Dir. 2018/2001/EU, Art. 3.2.

65　Regulation (EU) 2018/1999 of the European Parliament and of the Council of 11 December 2018 on the governance of the Energy and Climate Action Union, amending Regulations (EC) No 663/2009 and (EC) No 715/2009 of the European Parliament and of the Council, Directives 94/22/EC, 98/70/EC, 2009/31/EC, 2009/73/EC, 2010/31/EU, 2012/27/EU, and 2013/30/EU of the European Parliament and of the Council, Directives 2009/119/EC and (EU) 2015/652 of the Council, and repealing Regulation (EU) No 525/2013 of the European Parliament and of the Council, OJEU 2018, L. 328, 1-77. On this text, see P. Thieffry, 'Les instruments procéduraux en matière climatique: les mécanismes de gouvernance de l'union de l'énergie et de transparence-facilitation de l'accord de Paris' (2019) Énergie - Env. Infrastr. Dossier 23.

66　For a first application of this regulation, see Commission, 2020 State of the Energy Union Report pursuant to Regulation (EU) 2018/1999 on Governance of the Energy Union and Climate Action, COM (2020) 950 final, 14 October 2020.

with the major principles, loyal cooperation[67] or the principle of solidarity,[68] that govern relations between the Member States and the EU, and also extends it to regional scales, which may be broader (Article 12.7), and finally calls on the major figures of contemporary governance, participation (Articles 10 and 11) and public information (Articles 10 and 11 and Article 28).

1.4 Conclusion

If the directive (EU) 2018/2001 participates in the process of Europeanisation of the promotion of the use of RE, it also participates in its renewal by making evolve, in particular, the Europeanisation of the objectives of consumption of the said energy. Only the future will tell whether this new way of conceiving the Europeanisation of the promotion of RE, which is based on a genuinely collective commitment, is the best way to respond to the major issues on which it is based.[69]

Bibliography

1. Pierre Sabliere, Droit de l'énergie (Dalloz, 2013).

2. Fabienne Peraldi-Leneuf, "La politique européenne sur les ENR et l'efficacité énergétique : éclatement des responsabilités ou politique intelligente," in Bluman C. (ed.), Vers une politique européenne de l'énergie (Bruylant, 2012).

3. Corinne Lepage, 'Les différents accords sur l'énergie intervenus au niveau européen sont-ils suffisamment ambitieux?' [online] Available at: <https://www.actu-environnement.com/blogs/corinne-lepage/

---

67 This is a general principle of EU law which, in particular, requires that 'Member States shall take all appropriate measures, whether general or particular, to ensure fulfilment of the obligations arising out of the Treaties or resulting from action taken by the institutions' (TEU, Article 4.3).

68 In this spirit, the directive perpetuates the principle of exchanges between Member States of the results they obtain in terms of renewable energy consumption: see Dir. 2018/2001/EU, Art. 8.

69 It should be noted that, whatever its level, the result obtained by each State is calculated on the basis of data collected under Regulation (EC) No 1099/2008 of 22 October 2008 on energy statistics, supplemented by additional specific data transmitted to Eurostat by the national administrations.

77/corinne-lepage-accord-energie-europe-ambition-112.html> [Accessed 26 June 2018].

4. Sophie Fabregat, "Climat: le plan de la Commission pour réduire les émissions de carbone de 55 % d'ici 2030." [online] Available at: <https://www.actu-environnement.com/ae/news/Climat-union-europeenne-reduction-ges-transport-batiment-36122.php4> [Accessed 17 September 2020].

5. Von Der Leyen U., State of the Union Address, 16 Sept. 2020.

6. Angus Johnston, Eva van der Marel, "Ad Lucem? Interpreting the New EU Energy Provision, and in particular the Meaning of Article 194(2) TFEU," 22(5) European Energy and Environmental Law Review (2013), p. 181.

7. Frédérique Berrod and Antoine Ullestad, La mutation des frontières dans l'espace européen de l'énergie (Larcier, coll. Paradigme, 2016).

8. Patrick Thieffry, "Les instruments procéduraux en matière climatique: les mécanismes de gouvernance de l'union de l'énergie et de transparence-facilitation de l'accord de Paris," Énergie - Env. Infrastr. 2019, Dossier 23.

2

External Relations of The European Union in the Energy Sector: EU-Russia Energy (Trade) Relations

Rafael Leal-Arcas and Iuliia Dragunova

2.1 Introduction

The energy sector is a vital segment of any economy, be it a sovereign State, a political or an economic union of States. Hence, securing essential supply of energy, by internal or external means, is a crucial goal for any sovereign establishment. Complications come into play when several sovereigns act both in a united and individual capacity, depending on the context, as in the case of the European Union ("EU," the "Union").

This chapter aims to provide an analysis of the EU external relations in the energy sector, focusing on the EU-Russia energy trade relations and the Energy Community. The hypothesis we have undertaken is that securing reliable flows of diverse sources of renewable energy would foster the EU's energy security and decrease its vulnerability towards major fossil fuel exporters. There are currently efforts towards much greater investment in renewables in the EU and major oil companies will need to learn from renewables companies if they wish to enter this new field. In fact, the pandemic created by Covid-19 has offered the world what a renewable future might look like.

For instance, Rome-based Enel (which is the EU's largest utility) intends to invest €160bn by 2030 to triple its renewable-energy capacity and transform its grids in Europe and Latin America in order to be ready for an electric future.[70] Spain's Iberdrola, another very large energy company, has pledged to invest €75bn in renewables and grids by 2025.[71] The same green trend is true in other parts of the world outside the EU. For instance, even if under President Trump, the US became the world's larg-

---

70 Schumpeter, 'How Enel became Europe's climate centurion' *The Economist*, 28 November 2020, 64.
71 Ibid.

est producer of crude oil[72]—which does not mean that the US is energy independent[73]—America's NextEra, a pioneering utility, has also decided to invest heavily in wind and solar energy.[74] China is also investing heavily in renewables.[75] The scope of this chapter, however, is limited to the participation of the EU in the Energy Charter Treaty ("ECT"),[76] the EnC Treaty, and a general analysis of the EU-Russia energy trade relations.

After this short introduction, Section II provides an overview of pertinent data related to the EU's energy production and consumption, as well as current trends and challenges the EU faces in the energy realm. Section III covers related competence issues with regard to the EU treaties. Section IV examines the energy-related treaty regimes that the EU is a part of, whereas Section V analyses the EU's participation in the Energy Community ("EnC") and related issues. EU-Russia energy relations concerning trade in oil and gas are analyzed in Section VI, which is followed by the conclusion with recommendations in Section VII.

2.2 EU Energy Consumption and Production, Current Trends and Challenges

2.2.1 EU Consumption

The EU is one of the world's largest consumers of energy. In 2017,[77] the EU-27[78] held the third place in the total energy supply, the index which

---

72 Yet the US's oil boom is going down and coal-fired plants seem to be on their way out. See The Economist, 'What Donald Trump did for hydrocarbons' 24 October 2020, 34.
73 In 2019, the US imported 9.1m barrels a day of petroleum. Ibid.
74 Schumpeter, 64.
75 Ideally, many Asian countries that are not yet ready to make the move to renewable energy should develop natural gas instead of coal in the coming years. This move may not be the best option for climate-change mitigation, but it is a step in the right direction.
76 R. Leal-Arcas (ed.), *Commentary on The Energy Charter Treaty*, Edward Elgar Publishing Ltd, 2018.
77 For all statistical data provided in this section, data of 2017 will be used, unless explicitly stated otherwise. In 2017, Eurostat issued the relevant data on the EU-27, the EU-28 and energy.
78 For the purposes of this chapter, the EU-27 excludes the United Kingdom, which formally left the EU in 2020.

shows the amount of energy needed to satisfy total energy consumption demand of a particular geographical entity.[79] Petroleum products constituted 34.8% of the gross inland energy consumption, natural gas – 23.8%, solid fossil fuels – 13.6%. Hence, 72.2% of the EU-27 energy consumption was based on fossil fuels while allocating the remaining 12.6% and 13.9% of the gross inland energy consumption to nuclear heat and renewable energy sources respectively.[80]

Nevertheless, the distribution of shares of energy sources in gross energy consumption in different EU Members States depends on the economic structures, available natural resources, and national choices of particular countries.[81] It differs tremendously across the EU landscape.[82] For instance, the largest share of oil consumption was determined in Cyprus (92,5%) with it being a small island lacking natural resources and Luxembourg (63.5%) affected by 'fuel tourism.'[83] Renewable energy constituted more than 40% of energy sources in gross energy consumption only in Latvia (42.5%) and Sweden (41.5%), where in Luxembourg it amounted to merely 6.3%.[84] Natural gas share was fluctuating from 39.5% in the Netherlands to less than 2% in Sweden and Cyprus.[85]

---

79 Eurostat, *The EU in the World — 2020 edition* (Luxembourg: Publications Office of the European Union, 2020) <https://ec.europa.eu/eurostat/documents/3217494/10934584/KS-EX-20-001-EN-N.pdf/8ac3b640-0c7e-65e2-9f79-d03f00169e17> 119 (accessed 01 August 2020).

80 Eurostat, *Energy, Transport and Environment Statistics – 2019 Edition* (Luxembourg: Publications Office of the European Union, 2019) <https://ec.europa.eu/eurostat/documents/3217494/10165279/KS-DK-19-001-EN-N.pdf/76651a29-b817-eed4-f9f2-92bf692e1ed9> 20 (accessed 01 August 2020).

81 For further details on EU energy law and policy, see R. Leal-Arcas (ed.) *EU Energy Law and Policy: The external dimension*, Eliva Press, 2020; R. Leal-Arcas et al., *The Great Energy Transition in the European Union*, Volumes 1 and 2, Eliva Press, 2020; R. Leal-Arcas and J. Wouters (eds.), *Research Handbook on EU Energy Law and Policy*, Cheltenham: Edward Elgar Publishing Ltd, 2017; R. Leal-Arcas, C. Grasso and J. Alemany Rios, *Energy Security, Trade and the EU: Regional and International Perspectives*, Cheltenham: Edward Elgar Publishing Ltd, 2016; R. Leal-Arcas, *The European Energy Union: The Quest for Secure, Affordable and Sustainable Energy*, Claeys & Casteels Publishing, 2016; R. Leal-Arcas, A. Filis and E. Abu Gosh, *International Energy Governance: Selected Legal Issues*, Cheltenham: Edward Elgar Publishing Ltd, 2014.

82 Eurostat, *Energy Statistics*, 20.

83 Ibid 21.

84 Ibid.

85 Ibid.

In the evolution of the final (as opposed to gross inland index) energy consumption, the share of renewable energy sources has grown from 3.8% in 1990 to 9.7% in 2017.[86] The EU commitment to expand its share in its total energy consumption to 20% by 2020[87] seems to be an overestimation by now.

2.2.2 EU Production and International Trade Supply

In 2017, the EU's primary production of energy constituted only 44.9% of its total demand where the rest 55.1% have been secured from the external energy trade partners.[88] Renewables and waste energy sources constituted amounted to 29.9% of the primary energy production of the EU when around 27.8% was supplied by means of nuclear energy.[89] For earlier years, 66% of EU's primary production used to be attributed to just six member-states: the United Kingdom ("UK"), France, Germany, Poland and the Netherlands.[90] Over the decade of 2007-2017, the trend for fossil fuels primary energy production dropped tremendously while, on the opposite, a trend for renewable energy sources witnessed a 65.6% increase.[91] Nevertheless, although decreased, the 2017 EU total production amounted approximately to 4% share in petroleum and 13.6% share in natural gas, while being spread unequally among states.[92] It still required the gap between demand and supply to be closed by means of external trade relations.[93]

This data is illustrative of an idea of energy dependency where the shortfall of production has to be met by net imports. The EU-28 energy de-

---

86 Ibid 19.

87 European Commission, 'Communication from the Commission: Europe 2020. A European Strategy for Smart, Sustainable and Inclusive Growth' <https://eur-lex.europa.eu/legal-content/EN/TXT/PDF/?uri=CELEX:52010DC2020&from=en> 32 (accessed 18 December 2020). See also Rafael Leal-Arcas, 'Smart grids as a way to democratize energy' *Yale Journal of International Affairs*, 2020.

88 Eurostat, *Energy Statistics*, 19.

89 Eurostat, *The EU in the World*, 115.

90 *See* Rafael Leal-Arcas and Andrew Filis, 'Conceptualizing EU Energy Security Through an EU Constitutional Law Perspective' (2013) 36 *Fordham International Law Journal* 5, 1239.

91 Eurostat, *Energy Statistics*, 10.

92 Ibid., 11-12.

93 Eurostat, *The EU in the World*, 115.

pendency increased throughout last two and a half decades with its net imports of energy being larger than its primary production (from 44% in 1990 to 55.1% in 2017).[94] Since 2004, the EU's total dependency rate surpassed 50% rising to its highest in 2017 - 55.1% (with highest rates for oil (86.7%) and natural gas (74.3%)).[95] However, dependency rates with regards to various energy sources are also spread unevenly among the EU Member States. For instance, Denmark is entirely energy imports independent unlike Cyprus or Malta.[96] Although the EU has ambitious goals in switching to renewables (and is comparatively successful in doing so vis-a-vis other countries), this data reinforces the importance of the EU's external relations in the energy sectors, particularly those in energy trade, to fulfill its needs in fossil fuels demand. This raises another pertinent issue: the energy security of the Union.

2.2.3 Trends and Challenges in the EU External Relations in the Energy Sector

First and foremost, presented data delineates the disparity between the energy realities of EU Member States. Difference in energy mixes of Member States is mainly contingent on natural resources and industrial assets available, as well as political-economic relations with external energy suppliers. Such energy realities are crucial for lobbying the policies that separate members would aspire to promote at the EU level.[97] For instance, where Poland or the Baltic states might be opposed to intensifying the EU-Russia energy integration, Germany on Luxembourg could, on the contrary, pursue it. This leads to the lack of an always coherent stance with regards to the external energy relations of the Union from within of the establishment.

The EU is generally highly dependent on external energy supplies. The Commission recognises that the more globalised the energy market becomes, the more the EU increases its vulnerability towards global trends, e.g., price fluctuations or energy supply shocks.[98] Hence, pur-

---

94   Eurostat, *Energy Statistics*, 19.
95   Ibid.
96   *See* Leal-Arcas, 'Conceptualizing EU Energy Security' 1237.
97   Ibid 1241.
98   *See* Energy Roadmap, 2050.

suing strategic bilateral and regional external relations with energy suppliers reflect the EU's interests to ensure energy security and reliable supply flows.[99]

For decades, the European external energy policy has been primarily shaped by the legacies of single events, e.g., the 1973 oil crisis, the 2006 and 2009 Ukraine gas crises, the 2011 Fukushima nuclear incident, and the 2014 Russia-Ukraine crisis.[100] These events led to new forms of trade and new energy mixes.[101] However, in the last decades, the EU admitted the need for common policy. For example, in 2011, the EU Commission stated that the development of a coherent external EU energy policy would increase the collective EU energy security as well as assist the EU in addressing attempts of external suppliers to influence markets or impose its energy leverage in political realm.[102] In 2015, the Foreign Affairs Council adopted Council Conclusions on EU Energy Diplomacy that contained the EU Energy Diplomacy Action Plan.[103] Decarbonisation and diversification of the energy mix have been the central points of the EU energy policy that aims at promoting energy security, decreasing energy dependency and protecting the environment have been. Pursuant to the Title XXI of the Lisbon Treaty, the EU energy policy is clearly linked to the EU's environmental goals. However, this view has faced certain criticism of the conditionality of the EU energy policy.

To sum up, differences of the energy realities in the EU Member States bear considerable impact on the challenges that lay in front of the EU as a single Union to pursue common energy policy. The EU while being

---

99 See Rafael Leal-Arcas, Juan Alemany Rios and Costantino Grasso, 'The European Union and its energy security challenges' (2015) 8 *Journal of World Energy Law and Business*, 292.

100 See Luca Franza, Coby van der Linde and Pier Stapersma, 'The internal and external dynamics of EU energy relations.' (2018) 72 *Clingendael Spectator* 2. <https://spectator.clingendael.org/pub/2018/2/eu-energy-relations/> (accessed 01 August 2020).

101 Ibid.

102 European Commission, *An External Policy to Serve Europe's Energy Interests*. (2006). Paper from Commission/SG/HR for the European Council <https://www.consilium.europa.eu/ueDocs/cms_Data/docs/pressData/en/reports/90082.pdf> (accessed 01 August 2020).

103 European Union External Action, *EU Energy Diplomacy*, (2016). <https://eeas.europa.eu/topics/energy-diplomacy/406/eu-energy-diplomacy_en> (accessed 02 August 2020).

highly dependent on external energy supplies aspires to lower such dependence by means of pursuing energy security and "greening" of its total energy mix.

2.3 Competence Issues Regarding the EU Treaties

It has never been an easy task to achieve the common stance of the EU external action matters, even less so when they are related to energy. Discrepancies in geopolitical context, energy priorities and energy security capabilities make it more unrealistic to achieve common EU energy security, which is in the realm of the EU exclusive competence.[104] Moreover, energy is a multifaceted concept that lies across a variety of policy areas, such as foreign trade, human rights, foreign policy, national security, economics, etc.[105] This adds complications to the fact that the EU competences are unequally shared between the Union and its Member States depending on particular issue. Moreover, the EU competencies involve actions by a variety of agencies and institutions, with provisions enshrined across the Treaty on European Union ("TEU") and Treaty on the Functioning of the European Union ("TFEU").[106]

Considerable shift towards fulfilling energy policy and dividing competences has been achieved with the conclusion of the Treaty of Lisbon ("Lisbon"). Before Lisbon, energy was not a distinct area of foreign policy where Lisbon introduced a new energy section and expanded the foreign policy toolkit by establishing a European External Action Service.[107] The treaty enshrines a specific section on energy, which affirms most critical responsibilities and general energy-policy goals. Nevertheless, although the Treaty of Lisbon has reinforced the political value of energy, it has not assigned considerable additional decision-making power to the Union's institutions. This imposes limitations on the EU's ability to sustain coherent external energy policy.[108]

---

104 *See* Leal-Arcas, 'Conceptualizing EU Energy Security,' 1254.
105 Ibid 1255.
106 Ibid.
107 See Sascha Müller-Kraenner (ed.), 'The external relations of the EU in energy policy.' (2010) *Berlin: Heinrich Böll Foundation* <https://www.ecologic.eu/sites/files/publication/2015/external-relations-of-the-eu-in-energy-issues_0.pdf> (accessed 06 August 2020) 6.
108 *See* Müller-Kraenner, 6.

Although the EU has the institutional capacity to act as a single unit on behalf of all its Member States, it does not possess such capacity with regards to all policy areas, especially with regard to such complex one as energy.[109] The competences of the EU are shared between the Union and its Member States in the following ways: (1) the EU exclusive competences where the Union can exclusively act on behalf of member states to their exclusion; (2) shared competences where both the Union and its Member States have a degree of competence to act upon and usually relate to policy areas, for which treaties do not make express provisions; (3) exclusive competence of the EU Member States where the Union lacks any powers.[110]

As the EU is a customs union, questions related to external trade relations, common commercial policy, and competition lay in the realm of the EU exclusive competence. Hence, according to Article 3 of the TFEU, such energy-related matters as fostering competitive conditions of the energy trade in the internal market or addressing the issue of tariffs when energy commodities enter the EU, are within the exclusive competence of the EU.

Under Articles 4(2)(i) and 194 of the TFEU, energy per se (in its wide sense) lies within the realm of shared competencies between the EU and its members. Matters of shared competence may be handled to the EU pursuant to the subsidiarity and proportionality principles when the actions seem to be more efficient at the supranational level.[111] Hence, where the external energy-related policy matter falls within the first category, the EU would enjoy unrestrained powers to pursue it. When the latter category comes into play, it will be necessary to assess how the proposal should be decided at the EU level, unanimously or by majority for the policy to be adopted.[112]

Nevertheless, each EU member-state has the right to determine its own conditions for exploiting its energy resources and determining its energy mix.[113] The compelling energy security measures with distinct diplomatic

---

109 See Leal-Arcas, 'Conceptualizing EU Energy Security' 1227.
110 Articles 3, 4, 6 of the TFEU.
111 See Leal-Arcas, 'Conceptualizing EU Energy Security' 1252.
112 Ibid 1245.
113 Article 194(2) of the Lisbon Treaty.

nature remain within the powers of the EU Member States. This seems to be reasonable from the perspective of preserving certain of their sovereign prerogatives. In addition, implementation of energy-sector treaties concluded by the EU is also within the realm of exclusive competence of Member States.

The complications of the competence allocation are related to the lack of the clear-cut division of such competences. Moreover, different energy priorities and backgrounds of the EU Member States contribute to the difficulty to reach consensus. However, explicit inclusion of energy to the realm of competences along with further development of related policies delineate the steady development in the energy-related domain.

2.4 Participation in Treaty Regimes

The Treaty of Lisbon formalised the legal personality of the Union, so the EU became capable of concluding international agreements and joining international organisations as a single entity,[114] which is fundamental for engaging in external energy relations. Many EU Member States, separately and united, participate in various fora to promote their energy interests, e.g., the ECT,[115] the EnC, the North Atlantic Treaty Organisation ("NATO"), the International Energy Agency ("IEA"), the International Solar Alliance ("ISA"), the Organisation for Economic Cooperation and Development ("OECD"), the Paris Agreement framework, etc. The competence allocation also bears its impact on the conclusion of energy-related treaties.

The EU has bilateral and regional cooperation agreements with all of its neighboring countries. For instance, external energy relations with Norway, which is one of the most critical suppliers of oil and gas for the EU, are governed by the Agreement of the European Economic Area that makes Norway an integral part of the EU's internal market.[116] The European Energy Community Treaty was concluded with the EU's eastern and southern neighboring states to integrate them into the EU internal energy market (it will be analysed in detail in the next section). The EU concluded the Eastern Partnership association agreement with some of

---

114 Article 216 of the TFEU.
115 Italy withdrew in 2016.
116 *See* Müller-Kraenner, 4.

the CIS States to reduce high dependence on Russian energy imports, diversify fossil fuels suppliers and integrate these States in the EU's energy infrastructure. As for the Arctic's region, although Denmark is part of the Arctic Council, the EU does not participate in it sui juris.[117]

As for the multilateral efforts, the United Nations ("UN") Framework Convention on Climate Change and the Paris Agreement play a fundamental role in the EU securing the reduction of greenhouse gas ("GHG") emissions and increasing the share of renewables in the EU's energy mix.[118] The World Trade Organisation ("WTO") instruments enable the EU to ensure energy transit and various energy-related matters on the supranational level. The EU participates in the International Solar Alliance that promotes efficient exploitation of solar energy to reduce the share of fossil fuels in the energy mix.[119] Moreover, the EU participates in the International Renewable Energy Agency ("IRENA"). Although these multilateral initiatives are crucial for the EU's external policy relations, their analysis does not fall within the scope of the current paper. Therefore, we will focus in more detail on the ECT.

The ECT is the most frequently involved treaty in investor-state dispute settlement ("ISDS").[120] This treaty was signed in 1994 and demonstrated an attempt to apply WTO's free-trade disciplines to the trans-European

---

117 *See* Leal-Arcas, 'Conceptualizing EU Energy Security' 1275.

118 For further details, see Rafael Leal-Arcas *et al.*, 'Of international trade, climate change, investment and a prosperous future' (2020) 12 *Trade, Law and Development* 2; R. Leal-Arcas, 'Climate Clubs and International Trade across the European and International Landscape' (2020) 29 *European Energy and Environmental Law Review* 3, 72-88; R. Leal-Arcas et al., 'The contribution of free trade agreements and bilateral investment treaties to a sustainable future' (2020) 23 *Zeitschrift für Europarechtliche Studien – ZEuS* 1, 3-76; R. Leal-Arcas, *Solutions for Sustainability: How the International Trade, Energy and Climate Change Regimes Can Help* (Springer International Publishing 2019); W. Leal and R. Leal-Arcas (eds.), *University Initiatives in Climate Change Mitigation and Adaptation* (Springer International Publishing 2018); R. Leal-Arcas, *Climate Change and International Trade* (Cheltenham: Edward Elgar Publishing Ltd, 2013).

119 <https://thediplomat.com/2020/05/can-the-international-solar-alliance--truly-be-indias-gift-to-the-world/> (accessed 10 August 2020).

120 *See* Martin Dietrich Brauch, 'Modernizing the Energy Charter Treaty: A make--or-break moment for sustainable, climate-friendly energy policy' (2019) *International Institute for Sustainable Development Blog* <https://www.iisd.org/articles/modernizing-energy-charter-treaty-make-or-break-moment-sustainable-climate-friendly-energy> (accessed 10 August 2020).

energy sector.¹²¹ It came into effect in 1998 and was signed by 41 states including the EU. However, the treaty has not been ratified by the major energy moguls, such as Russia or Norway, with the former suspending the treaty in 2009 after facing the Yukos saga. The ECT contains disciplines on energy-sector foreign investment and its protection once such investment has been admitted to the host state territory. Non-discriminatory treatment of investments (most-favoured nation and national treatment standards) is the foundation of the Treaty.

The ECT has been continuously subject to criticism due to undermining the environmental protection initiatives, such as the Green Deal and the Paris Agreement, and allegedly being a vehicle for fossil fuel industries as it hinders states to issue green policies and raise public interest related defences in ISDS.[122] Renewable energy generally received little attention when the ECT was drafted. For various reasons, the call for its modernisation has been roaring throughout the last decade.[123]

In 2012, parties to the treaty agreed to commence the process of modernisation during negotiations rounds called the Warsaw Process. However, the Energy Charter Conference, an intergovernmental decision-making body for the ECT process, made an official announcement of beginning of these renegotiation efforts only in 2019 while still lacking firm consensus among its members on the direction of the amendments, especially in relation to investment protection and dispute settlement.[124] The latest (third) round of negotiations took place in November 2020.[125] On 2 December 2020, the representative to the European Commission stated that "[i]f core EU objectives ... are not attained within a reasonable timeframe, the Commission may consider proposing other options, including the withdrawal from the ECT."[126]

---

121   *See* Müller-Kraenner, 4.
122   *See* Brauch.
123   Ibid.
124   Ibid.
125   European Commission, 'Third negotiation round to modernise Energy Charter,' (2020) *European Commission Website*, <https://trade.ec.europa.eu/doclib/press/index.cfm?id=2206> (accessed 18 December 2020).
126   'Answer given by Executive Vice-President Dombrovskis on behalf of the European Commission,' P-005555/2020, 2 December 2020, <https://www.europarl.europa.eu/doceo/document/P-9-2020-005555-ASW_EN.pdf> (accessed

In light of the global commitment to climate action, the ECT needs a "fundamental overhaul" with the new version encouraging low-carbon investments and discouraging fossil fuels ones, supporting policy measures to accommodate transition to the green economy.[127] It is particularly critical for the EU while being a pioneer in the environmental march. In case such treaty modernisation ceases to be feasible, the EU would indeed most probably have to withdraw from it, as the treaty as it is today opposes the EU's aspirations in the energy sector.

Generally, energy-related treaty regimes represent a critical instrument for the EU in achieving its goals related to decarbonisation and energy security, decreasing energy dependency on major fossil fuels suppliers.

2.5 Participation in the Energy Community

The EnC was set up by the Energy Community Treaty ("EnC Treaty") between the EU and several neighboring states in 2005. In 2006, it entered into force and, in 2016, was extended until 2026.[128] It aims at expanding the EU internal energy market to its neighboring countries while creating a stable regulatory and market framework to attract investment, creating integrated regional market related to the EU and its regulations, ensuring security of supply, boosting competition and fostering environmental protection.[129] The EnC framework obliges its participants to adopt the key EU energy laws. Eventually, the EnC has established its own body of law known as EnC acquis that has to be transported in domestic legislations of member states.[130] Hence, its crucial role is to ensure regulatory convergence to achieve a more coherent and predictable

---

18 December 2020).

127 *See* Brauch.

128 European Parliament, *Energy Community: Prospects and Challenges*, (2015) Briefing, <https://www.europarl.europa.eu/RegData/etudes/BRIE/2015/569011/EPRS_BRI(2015)569011_EN.pdf> (accessed 10 August 2020).

129 Article 2 of the EnC Treaty. See more generally R. Leal-Arcas, *International Trade and Investment Law: Multilateral, Regional and Bilateral Governance* (Cheltenham: Edward Elgar Publishing Ltd 2010); R. Leal-Arcas (ed.) *The Future of International Economic Law and the Rule of Law* (Eliva Press 2020); R. Leal-Arcas (ed.) *International Trade, Investment and The Rule of Law* (Eliva Press 2020).

130 European Parliament, *Energy Community*, 2.

investment environment and more integrated energy market for the EU and its neighbors.[131]

Initially, the treaty was concluded by the potential EU-membership candidate countries (Albania, Bosnia-Herzegovina, Kosovo, Macedonia, Montenegro, and Serbia). However, accession of Moldova (2010) and Ukraine (2011) expanded the geo-political perspective of the treaty to non-candidates. Generally, membership is open to other States. Several parties to the treaty eventually joined the EU, e.g., Bulgaria, Romania, and Croatia. Non-EU states can participate as observers, e.g., Georgia, Armenia, Norway, and Turkey. The European Commission primarily represents the EU in its interaction with the EnC parties. However, the EU Member States can still attain a status of a participant on its own behalf, which has been done by 19 EU members.

Although the EU strongly supports the EnC objectives, the shortcomings have been identified in the EnC framework, e.g., issue of potential expansion, weak implementation of the energy acquis in several state-parties, lack of adequate mechanisms for enforcement. These issues have been commissioned to be analysed by the specially organised high-level group and addressed in the following Ministerial Councils.[132]

Though initially created to accommodate Balkan states as potential EU membership candidates, the EnC framework has expanded beyond such potential members participation. This draws to the question of whether the single membership is realistic within the EnC context when its parties are not adopting the acquis with the same rigor, which weakens the credibility of the establishment. Although widening the EnC membership might serve the geopolitical interests of the EU, the potential EnC members should possess a strong incentive to implement all necessary regulations, which might be too burdensome for states lacking a realistic prospect of the EU membership.[133]

The initial EU acquis package included provisions on matters related to gas and electricity markets, environment, renewable energy, obligations to sign the Kyoto Protocol, to apply the EU principles in the areas of com-

---

131 *See* Leal-Arcas, 'Conceptualizing EU Energy Security' 1261-62.
132 European Parliament, *Energy Community*, 1.
133 Ibid.

petition, etc. The EnC includes a rather flexible mechanism allowing the acquis to expand to new areas of the EU law making it dynamic and reflective of changing reality. Although "dynamic" acquis allows the EnC to rapidly adapt to changing EU law, in many cases, parties to the treaty merely transplant the regulations into their domestic legislation without proper implementation.[134]

Albeit the fact that this treaty envisages tools for enforcement, e.g., a case can be taken to the Secretariat by a public authority, legal entity or person, so far, with 19 enforcement cases commenced, binding penalties and financial sanctions have not been introduced. Two of these cases resulted in breach of the EnC Treaty, however, nothing sufficient has been done from the side of the State parties to remedy the situation and implement enforcement decisions. Hence, lack of binding financial sanctions leads to fundamental enforcement problems.

The EnC framework illustrated an adequate tool for the EU to strengthen its external energy policy and ensure energy security and diversification of energy sources. However, the regime requires fundamental revision and modernisation. Certain recommendations on such changes have been made by the High-Level Reflection Group in its assessment of the efficiency of the EnC, e.g., establishment of the independent Court of Justice, ensuring direct effect of the EnC Treaty in domestic legislations of parties to the treaty, providing a flexible two-tier membership. However, the following public consultations lacked the consensus on these recommendations. For now, many potent initiatives have been undertaken under the auspices of EnC, such as Energy Community Regulatory School increasing expertise of its members authorities or the dispute settlement center resolving ISDS disputes. Hence, the EU does not cease to grow its external energy policy leverage and influence among neighboring states.

---

134   European Parliament, *Energy Community*, 4-5.

Chapter 2

2.6 EU-Russia Trade Relations in the Energy Sector[135]

As a result of globalization, real GDP per capita in emerging economies more than doubled between 1995 and 2019 in purchasing-power-parity terms.[136] In the case of developed economies, however, it grew only by 44%.[137] Between 1990 and 2008, international trade as part of global GDP rose from 39% to 61%.[138] Explosive trade growth, however, has come to an end,[139] the end of hyperglobalization seems to have arrived, trade has started to go regional, globalization has become slowbalization, and openness in global trade seems to have also come to an end, with the protection of domestic producers as a matter of national security. Moreover, global competition has placed much pressure on exporters to become more efficient and competitive.

In the specific case of energy trade relations, Russia is the largest supplier of oil, natural gas, and hard coal to the EU.[140] In 2017, 60% of total imports from Russia to the EU were in the energy sector. Considering the large gap between the primary energy production of the EU in the realm of fossil fuels and the pertinent demand, the relationship with Russia still remains as one of the most critical in the energy sector.

The EU-Russia energy relationship was strengthened during the aftermath of the Iraq War in 2003, when the EU turned closer to Russia to secure its energy supply while turning away from US leadership.[141] For

---

135 For a detailed analysis of how EU trade law and policy function, see R. Leal-Arcas, *Eu Trade Law* (Elgar European Law Series, Edward Elgar Publishing Ltd 2019); R. Leal-Arcas, *Theory and Practice of EC External Trade Law and Policy* (London: Cameron May 2008).

136 The Economist, 'Free exchange (Turning inward): Why is the idea of import substitution being revived?' 7 November 2020, 64.

137 Ibid.

138 Ibid.

139 That said, trade has not done as poorly as initially thought. In April 2020, the World Trade Organization predicted that goods trade would fall by 13-32% in 2020; in October 2020, it seemed to be only around 10%. See *The Economist*, 'Special report: The world economy' 10 October 2020, 1-14, at 8.

140 Eurostat, 2017.

141 *See* Isabelle Beegle-Levin, 'EU-Russia Energy Relations: Considerations for United States Policymakers,' (2018), *Master Thesis* <https://www.researchgate.net/publication/326426407> (accessed 13 August 2020).

Russia, with the global oil prices raising tremendously at that moment, the EU market was an ideal consumer. The 2008 Russia-Georgia conflict made the EU more vigorously engage in diversifying its energy mix to shift from its compelling energy dependency on the oil-rich partner. The 2014 Russia-Ukraine crisis reiterated the EU's aspirations towards decreasing overreliance on Russia's energy resources, which the latter has been continuously using as a geopolitical leverage.[142]

The EU-Russia external energy relations are largely sustained by means of bilateral and (less) multilateral efforts. However, the institutionalised regimes between two entities, such as the ECT and the 1994 Partnership and Cooperation Agreement, for now, ceased to exist for various reasons. The latter agreement has governed the relations for a decade after its ratification in 1997 and, for that period, has been the main legal framework for EU-Russian energy trade and investment. Nevertheless, the fact that these frameworks ceased to exist has not discouraged mutual trade and investment flows.[143] The European Commission still perceives Russia as an important energy partner, regardless of the strong EU's commitment to decarbonisation.[144]

The relationship with Russia is central to the external energy policy of the EU, being its *raison d'etre*.[145] However, it would be deceitful to perceive energy as purely economic matter when entrusted to a supranational body, such as the EU, where its Member State can separately pursue its energy needs and interests.[146] The approach towards this relationship differs across the EU Member States. For instance, Germany's energy relations with Russia are more favourable than those with Hungary's (or any other state with difficult historic relations with Russia and its predecessors, the USSR and the Russian Empire). Hence, Germany will not only seek to ensure that its energy interests are well served on the national level but also attempt to secure a beneficial framework with Russia on the level of

---

142   Ibid.
143   *See* Leal-Arcas, 'Conceptualizing EU Energy Security' 1288-89.
144   Ibid., 1266-67.
145   Oliver Geden, Clémence Marcelis, and Andreas Maurer, 'Perspectives for the European Union's External Energy Policy' (2006), *Working Paper FG 1, 2006/17 SWP Berlin* <https://www.swp-berlin.org/fileadmin/contents/products/arbeitspapiere/External_KS_Energy_Policy__Dez_OG_.pdf> 16.
146   *See* Leal-Arcas, 'Conceptualizing EU Energy Security' 1271.

the EU. Construction of the Nord Stream 1 and 2 pipelines and circumvention of Russia's neighbors with strained relations along the way with direct supply to Germany represent an evident illustration of disparity in national interests of the EU Member States when it comes to the question of energy relations with Russia.[147]

However, Russia desires a good relationship with the EU energy market as much as the EU is reliant on Russia's reliable supplies of natural gas and oil. Russia's energy policy perceives the European market as vital in the energy sector.[148] Russia is interested in a stable demand for its energy resources, ensuring the economic efficiency of the activities of Russian energy companies, attracting investments, advanced technologies and experience.[149] For its part, Russia is trying to ensure guarantees of stability and reliability of supplies, flexibility in pricing the proposed energy resources, technological interaction with European energy systems. Therefore, the central aspect of the EU-Russian external energy relations is interdependency.

Nevertheless, Russia has used its ascendancy in the energy sector as a leverage in its geopolitical aspirations, e.g., in the Ukrainian gas crisis of 2000s, which led the EU to pursue a more independent energy mix as well as more secure supply. Certain energy policy interests might differ substantially between Russia and the EU. The EU while concerned about supply of the required amount of goods to its internal market promotes energy market liberalisation, decrease of energy subsidies, foreign investment protection. Russia, on the contrary, is interested in keeping as much control over its energy system and its influence as possible, securing constant flow of hard currency needed to upgrade its upstream and downstream infrastructure.[150]

To sum up, the EU is already successfully addressing mentioned issues by adopting strong policies towards low-carbon energy sources supply,

---

147 See Franza et al.
148 Ivanov and Likhachev, 'Relations of Russia and the EU in the Energy Sector,' (2017) *Russian Council on International Affairs* publication <https://russiancouncil.ru/papers/Russia-EU-Energy-RIAC-DGAP-Report35ru.pdf> (accessed 13 August 2020) 13.
149 Ibid.
150 See Leal-Arcas, 'Conceptualizing EU Energy Security' 1295-96.

diversifying its energy mix and energy suppliers.[151] The fact that some EU Member States perceive Russia as a "whimsical" partner where others consider it to be a reliable ally[152] complicates achieving the common external energy strategy towards Russia. What can be said without hesitation is that the EU and Russia energy relations will continue to develop with the EU's high energy demand and Russia's possibilities of supply, hopefully, along the common track of shifting to a greener economy.

2.7 Conclusion

Energy is a vital part of the existence of any social or economic system, be it a state or a State union. The EU energy structure is highly dependent on external supplies. This fact explains why external energy relations play a crucial role in foreign-policy development. The ability to accommodate an effective EU energy acquis among its neighboring states as well as ensure a stable relation with the biggest energy supplier are two of the major external goals of the Union. However, to do so efficiently while ensuring energy security requires certain policy solutions, which the EU has been implementing for some time: diversifying energy supplies, e.g., by means of trade liberalisation in green goods and services, the promotion of renewables, as well as the lobbying of the ECT modernisation. Finally, securing reliable flows of diverse and sustainable energy sources would foster the EU's energy security.

Bibliography

1. Eurostat, The EU in the world — 2020 edition (Luxembourg: Publications Office of the European Union, 2020) <https://ec.europa.eu/eurostat/documents/3217494/10934584/KS-EX-20-001-EN-N.pdf/8ac3b640-0c7e-65e2-9f79-d03f00169e17> 119 (accessed 01.08.2020).

2. Eurostat, Energy, Transport and Environment Statistics – 2019 Edition (Luxembourg: Publications Office of the European Union, 2019) <https://ec.europa.eu/eurostat/documents/3217494/10165279/KS-DK-19-001-EN-N.pdf/76651a29-b817-eed4-f9f2-

---

151 *See* Beegle-Levin, 82.
152 *See* Leal-Arcas, 'Conceptualizing EU Energy Security' (n 151).

92bf692e1ed9> 20 (accessed 01.08.2020).

3. Energy Roadmap 2050.

4. European Commission, An External Policy to Serve Europe's Energy Interests (2006) Paper from Commission/SG/HR for the European Council <https://www.consilium.europa.eu/ueDocs/cms_Data/docs/pressData/en/reports/90082.pdf> (accessed 01.08.2020).

5. European Union External Action, EU Energy Diplomacy (2016) <https://eeas.europa.eu/topics/energy-diplomacy/406/eu-energy-diplomacy_en> (accessed 02.08.2020).

6. The Treaty on Functioning of the European Union.

7. The Lisbon Treaty.

8. The Treaty on the European Union.

9. The European Energy Community Treaty.

10. High Level Reflection Group of the Energy Community, 'An Energy Community for the Future' (2014) Energy Community Report.

11. European Parliament, Energy Community: Prospects and Challenges (2015) Briefing <https://www.europarl.europa.eu/RegData/etudes/BRIE/2015/569011/EPRS_BRI(2015)569011_EN.pdf> (accessed 10.08.2020) 2.

12. Rafael Leal-Arcas and Andrew Filis, 'Conceptualizing EU Energy Security Through an EU Constitutional Law Perspective' (2013) 36 Fordham International Law Journal 5, 1239.

13. Rafael Leal-Arcas, Juan Alemany Rios and Costantino Grasso, 'The European Union and its energy security challenges' (2015) 8 Journal of World Energy Law and Business 292.

14. Luca Franza, Coby van der Linde and Pier Stapersma, 'The internal and external dynamics of EU energy relations' (2018) 72 Clingendael Spectator 2 <https://spectator.clingendael.org/pub/2018/2/eu-energy-relations/>.

15. Sascha Müller-Kraenner, 'The external relations of the EU in energy policy' (2010) Berlin: Heinrich Böll Foundation (ed.) <https://

www.ecologic.eu/sites/files/publication/2015/external-relations-of-the-eu-in-energy-issues_0.pd> (accessed 0.08.2020) 6.

16. Martin Dietrich Brauch, 'Modernizing the Energy Charter Treaty: A make-or-break moment for sustainable, climate-friendly energy policy' (2019) International Institute for Sustainable Development Blog <https://www.iisd.org/articles/modernizing-energy-charter-treaty-make-or-break-moment-sustainable-climate-friendly-energy> (accessed 10.08.2020).

17. Isabelle Beegle-Levin, 'EU-Russia Energy Relations: Considerations for United States Policymakers' (2018) Master Thesis <https://www.researchgate.net/publication/326426407> (accessed 13.08.2020).

18. Oliver Geden, Clémence Marcelis,* and Andreas Maurer, 'Perspectives for the European Union's External Energy Policy' (2006) Working Paper FG 1, 2006/17 SWP Berlin <https://www.swp-berlin.org/fileadmin/contents/products/arbeitspapiere/External_KS_Energy_Policy__Dez_OG_.pdf> 16.

19. Ivanov and Likhachev, 'Relations of Russia and the EU in the Energy Sector' (2017) Russian Council on International Affairs publication <https://russiancouncil.ru/papers/Russia-EU-Energy-RIAC-DGAP-Report35ru.pdf> 13.

20. Stephen Padgett, 'Multilateral institutions, accession conditionality and rule transfer in the European Union: the Energy Community in South East Europe' (2012) 32 Journal of Public Policy 261.

21. Roman Petrov, 'Energy Community As a Promoter of the European Union's 'Energy Acquis' to its Neighbourhood' (2012) 39 Legal Issues of Economic Integration 331.

22. Marco Siddi, 'The EU's gas relationship with Russia: solving current disputes and strengthening energy security' (2017) 15 Asia Europe Journal 107.

23. Matus Mišík, 'On the way towards the Energy Union: Position of Austria, the Czech Republic and Slovakia towards external energy security integration' (2016) 111 Energy 68.

24. Angelique Palle, 'Regional Dimensions to Europe's Energy Integra-

tion' (2013) The Oxford Institute for Energy Studies 9.

25. Jonathan Stern, 'Reducing European Dependence on Russian Gas: Distinguishing Natural Gas Security from Geopolitics' (2014) The Oxford Institute for Energy Studies 29.

3

THE EUROPEAN UNION AND THE LEGAL GOVERNANCE OF OFFSHORE ENERGY ACTIVITIES

Nathalie Ros

3.1 Introduction

Energy is at the heart of the historic dimension of European integration, now illustrated by the Energy Union project, initiated by the Commission in 2015, in connection with an ambitious EU climate policy.[153] Nevertheless, the EU energy dependency is increasing, though it is naturally variable from one State to another. More than half (55.1 % according to Eurostat)[154] of the EU energy consumption is imported, in particular hydrocarbons, i.e., oil and gas; this situation is not without geopolitical risks for the third largest energy consumer in the world.[155] The Union produces only 25.7 % of the gas and 13.7 % of the oil it consumes; its primary production has been very much in decline for more than ten years,[156] and it is therefore also necessarily for hydrocarbons that energy dependency rates are the highest.[157] In fact, the European Union's oil and gas reserves are modest; Norway is by far the largest producer in Europe, but it is only a member of the European Economic Area ("EEA"), and the largest producer in the Union was the United Kingdom, which means the Brexit is going to increase the EU dependency. The main European basins are located offshore, primarily in the North Sea, but also in the Atlantic,

---

153 Energy Union Package, Communication from the Commission to the European Parliament, the Council, the European Economic and Social Committee, the Committee of the Regions and the European Investment Bank, A Framework Strategy for a Resilient Energy Union with a Forward-Looking Climate Change Policy, Brussels, 25.2.2015, COM(2015) 80 final <https://eur-lex.europa.eu/resource.html?uri=cellar:1bd46c90-bdd4-11e4-bbe1-01aa75ed71a1.0001.03/DOC_1&format=PDF> (accessed 10 October 2020).

154 All the figures cited are taken from this document. Eurostat Statistics Explained, Energy production and imports, June 2019 <https://ec.europa.eu/eurostat/statistics-explained/index.php/Energy_production_and_imports> (accessed 10 October 2020).

155 After China and the United States, and with Russia and the Middle East as main suppliers outside the EEA.

156 From 2007 to 2017, primary production of hydrocarbons fell, with a reduction of 39.4 % for natural gas and 38.9 % for crude oil.

157 Namely 86.7 % for crude oil, and 74.3 % for natural gas.

the Baltic, the Black Sea, the Mediterranean, and even in the Norwegian Sea, the Barents Sea, or the Arctic. More than 90 % of oil production and 60 % of gas production in the EEA come from offshore deposits; and it is also at sea that the reserves are by far the most promising.

The European Union's interest in offshore activities is therefore obvious and a priori participates in a sustainable development approach: the EU has a strong interest in maintaining its own oil and gas production, not only to avoid further increasing its energy dependency, but also to ensure that the European economy can maintain jobs and related business opportunities; but the disaster occurred on 20 April 2010 on the Deepwater Horizon platform, operating the Macondo well in the Gulf of Mexico, also confronted Europe with the environmental and human dangers of this type of industrial activity, developed under increasingly extreme conditions, both in its own waters and in neighboring waters. Technological progress has indeed made it possible to push the limits of offshore exploitation to increasingly impressive depths,[158] but the dangerousness of such operations has naturally increased, a fortiori when the deposits are located in fragile seas due to their configuration, such as enclosed or semi-enclosed seas, like the Black Sea or the Mediterranean, or due to extreme climatic conditions such as in the Arctic Ocean.

In these circumstances, the Deepwater Horizon accident could therefore only lead the European Union to a kind of self-examination and an urgent assessment of the applicable legal framework. The Commission launched this study as early as May 2010, and a review of EU legislation on the safety of offshore oil and gas exploration and production activities in European waters was undertaken, as well as a consultation with industry and Member States competent authorities. Already in July 2010, this ultra-reactive strategy paid off by enabling the Commission to identify five main areas where action was needed to maintain the safety and environmental credentials of the EU: thorough licensing procedures, improved controls

---

158   The 1,000 meters depth record was set in Brazil in 1994, and the 2,000 meters in the Gulf of Mexico in 2002; on 20 February 2013, the world leader in offshore drilling *Transocean* announced a new depth record underwater with a wellhead of 3,165 meters in the Bay of Bengal. Drilling of more than 10 kilometers can now be carried out, the most infamous of which was operated on 2 September 2009 from the Deepwater Horizon platform, with a depth of 10,685 km including 1,259 meters of water, on the *Tiber* deposit in the EEZ of Mexico.

by public authorities, addressing gaps in applicable legislation, reinforced EU disaster response, and international cooperation to promote offshore safety and response capabilities worldwide.

The approach advocated by the European Union, and at least initially by the Commission[159] and the Parliament,[160] could therefore only be understood by reference to the Integrated Maritime Policy,[161] which precisely aims to create a framework that facilitates the development and coordination of the various maritime activities, in order to achieve better economic and human benefits with fewer negative effects on the environment, i.e., to maximize the sustainable exploitation of the seas and oceans, particularly in the very current prospect of the Blue Economy and Blue Growth,[162]

---

159 Communication from the Commission to the European Parliament and the Council, Facing the challenge of the safety of offshore oil and gas activities, Brussels, 12.10.2010, COM(2010) 560 final {SEC(2010) 1193 final} <https://eur-lex.europa.eu/LexUriServ/LexUriServ.do?uri=COM:2010:0560:FIN:EN:PDF> (accessed 10 October 2020).

160 P7_TA(2010)0352, EU action on oil exploration and extraction in Europe, European Parliament resolution of 7 October 2010 on EU action on oil exploration and extraction in Europe <http://www.europarl.europa.eu/sides/getDoc.do?pubRef=-//EP//NONSGML+TA+P7-TA-2010-0352+0+DOC+PDF+V0//EN> (accessed 10 October 2020); P7_TA(2011)0366, Safety of offshore oil and gas activities, European Parliament resolution of 13 September 2011 on facing the challenges of the safety of offshore oil and gas activities (2011/2072(INI)) <http://www.europarl.europa.eu/sides/getDoc.do?pubRef=-//EP//NONSGML+TA+P7-TA-2011-0366+0+DOC+PDF+V0//EN> (accessed 10 October 2020).

161 On the Integrated Maritime Policy in general, as well as on various sectoral and regional aspects, see Laura Carballo Piñeiro (ed.), *Retos presentes y futuros de la Política Marítima Integrada de la Unión Europea* (Editorial Bosch 2017); Timo Koivurova, 'A Note on the European Union's Integrated Maritime Policy' [2009] *Ocean Development & International Law* 171; José Manuel Sobrino Heredia, 'La protección marítima, nueva dimensión de la Política Marítima de la Unión Europea' [2007] *Revista de Derecho Comunitario Europeo* 417; José Manuel Sobrino Heredia, 'La politique maritime intégrée de l'Union européenne et les bassins maritimes européens' [2013] *Paix et Sécurité Internationales* 13; José Manuel Sobrino Heredia, 'Quelles 'déclinaisons méditerranéennes' de la stratégie maritime intégrée de l'Union européenne?' in Nathalie Ros et Florence Galletti (eds.), *Le droit de la mer face aux 'Méditerranées.' Quelle contribution de la Méditerranée et des mers semi-fermées au développement du droit international de la mer?* Cahiers de l'Association internationale du Droit de la Mer 5 (Editoriale Scientifica 2016) 85.

162 Communication from the Commission to the European Parliament, the Council, the European Economic and Social Committee and the Committee of the

which naturally includes offshore operations.[163] But the Union rapidly found itself confronted with a certain number of difficulties, resulting from the economic weight of the extractive industries, already very well established in some Member States, with very active lobbies, but also because of the specific nature of the offshore activity and the necessarily transboundary dimension of its dangers. Under these conditions, its effective contribution may indeed be limited in practice, both in scope and effects, particularly in the case of operations carried out in the five "shared sea-basins,"[164] and a fortiori outside EU waters but in neighboring waters, i.e., in marine regions which are not an integral part of the EU but are close to it.

The development of offshore oil and gas operations, in marine waters under the sovereignty and jurisdiction of European Union Member States, but also in neighboring waters, has therefore imposed the subsequent need for better regulation and supervision. The strategy of the European Union is therefore defined and implemented according to a twofold logic, which conforms to the binary dimension of its "policies and internal actions"[165] and its "external action."[166] It leads the Union to develop a dedicated leg-

---

Regions, Blue Growth: opportunities for marine and maritime sustainable growth, Brussels, 13.9.2012, COM(2012) 494 final <https://ec.europa.eu/maritimeaffairs/sites/maritimeaffairs/files/docs/body/com_2012_494_en.pdf> (accessed 10 October 2020); on legal issues related to Blue Growth, see Jaime Cabeza Pereiro y Belén Fernández Docampo (eds.), *Estrategia Blue Growth y Derecho del Mar* (Editorial Bomarzo 2018).

163 Nathalie Ros, 'Retos de la Política Marítima Integrada de la Unión Europea frente a las perspectivas de explotación mar adentro' in Laura Carballo Piñeiro (ed.), *Retos presentes y futuros de la Política Marítima Integrada de la Unión Europea* (Editorial Bosch 2017) 329.

164 The five shared sea-basins are the Arctic, the Atlantic, the Baltic, the Mediterranean and the Black Sea. Communication from the Commission to the European Parliament, the Council, the European Economic and Social Committee and the Committee of the Regions, Developing the international dimension of the Integrated Maritime Policy of the European Union, Brussels, 15.10.2009, COM(2009)536 final <http://www.europarl.europa.eu/RegData/docs_autres_institutions/commission_europeenne/com/2009/0536/COM_COM(2009)0536_EN.pdf> (accessed 10 October 2020).

165 By reference to Part Three of the Treaty on the Functioning of the European Union, *Union policies and internal actions*.

166 To use the terminology of Part Five of the TFEU, *External action by the Union*, which is defined *in fine* by Article 205 as 'The Union's action on the international scene' in other words the action of the EU acting as a subject of Public Interna-

islative framework that is as comprehensive and efficient as possible, but also to pursue its international cooperation efforts in favor of the legal regulation of offshore activities, if not effectively worldwide, at least beyond its borders, and in the regions and sea-basins it shares with non-Member States.[167] The achievement is a new EU strategy for the legal governance of offshore energy activities, developed from Cooperation in Public International Law to Integration into European Union Law.

3.2 From Cooperation in Public International Law

Outside its own legal order, it is essentially within the framework of the regional conventions and systems in which it participates, or is associated, that the European Union is confronted with the challenges of offshore, generally in a Material and/or Formal Soft Law Framework, but also in a Hard Law Framework: the Exception of the Mediterranean being from this point of view unique and all the more interesting.

3.2.1 In a Material and/or Formal Soft Law Framework

Depending on the situation, the Union is in fact involved in the management of these issues, both economic and environmental, either indirectly in those Regional Systems where the EU is not a Party, or directly when the UE is a Party to Regional Treaties.

3.2.1.1 Regional Systems where the EU is not a Party

This is the case in two of the five "shared sea-basins"[168] of the EU,[169] and

---

tional Law.
167 Nathalie Ros, 'L'Union européenne et l'encadrement juridique des activités offshore' (2018) *Revue du droit de l'Union européenne*, 233.
168 Communication from the Commission to the European Parliament, the Council, the European Economic and Social Committee and the Committee of the Regions, Developing the international dimension of the Integrated Maritime Policy of the European Union, Brussels, 15.10.2009, COM(2009)536 final <http://www.europarl.europa.eu/RegData/docs_autres_institutions/commission_europeenne/com/2009/0536/COM_COM(2009)0536_EN.pdf> (accessed 10 October 2020).
169 On this issue, and specifically with regard to the Black Sea and the Arctic, to which the study is devoted, Ioannis Stribis, 'Challenges in the implementation of the EU's Integrated Maritime Policy in shared sea-basins' in Laura Carballo Piñeiro (ed.), *Retos presentes y futuros de la Política Marítima Integrada de la*

first of all in the Arctic, although it does not *stricto sensu* concern activities developed in European waters, but this legal situation also exists in one of the four marine regions of the Union, in the Black Sea.

3.2.1.1.1 In the Arctic

Arctic waters are not part of the European Union's marine waters but they "are a neighbouring marine environment of particular importance for the Union";[170] this is justified for economic, strategic and environmental reasons, but also for their important role in mitigating climate change. Oil and gas industry activities are a reality[171] and a prospect in the Arctic, not only because the resources and reserves are promising, but also because their exploitation is likely to become less and less difficult, due to warming and melting ice.[172] In 2008, the European Parliament stated that "the Arctic region may contain approximately 20 % of the world's undiscovered oil and gas reserves."[173]

From an institutional point of view, the European Union cannot be a member of the Arctic Council and has failed to obtain permanent observer status, while three EU Member States are full members (Denmark, Finland, Sweden) and six (plus the United Kingdom) participate as observers (Germany, France, Italy, Netherlands, Poland, and Spain). In spite of this, the European Union's commitment to be involved has never wavered, and its strategy has resulted in the adoption of a certain number of texts dedicated to the Arctic and its governance.[174] Of course, the stakes

---

Unión Europea (Editorial Bosch 2017), 355.

170 Directive 2013/30/EU of the European Parliament and of the Council of 12 June 2013 on safety of offshore oil and gas operations and amending Directive 2004/35/EC, Official Journal of the European Union, 28.6.2013, L 178/72, Recital 52 <https://eur-lex.europa.eu/legal-content/EN/TXT/PDF/?uri=CELEX:32013L0030&from=EN> (accessed 10 October 2020).

171 In Alaska, but also in the Barents Sea.

172 Diethard Mager, 'Climate Change, Conflicts and Cooperation in the Arctic: Easier Access to Hydrocarbons and Mineral Resources?' (2009) *The International Journal of Marine and Coastal Law*, 347.

173 P6_TA(2008)0474, European Parliament resolution of 9 October 2008 on Arctic governance, Point H <http://www.europarl.europa.eu/sides/getDoc.do?pubRef=-//EP//TEXT+TA+P6-TA-2008-0474+0+DOC+XML+V0//EN> (accessed 10 October 2020).

174 The latest is the Joint Communication from the European Commission and the High Representative of the Union for Foreign Affairs and Security Policy to the

and risks associated with offshore are a large part of this interest.[175]

Thus, the European Parliament Resolution of 9 October 2008 on Arctic governance highlights the role of the Arctic region in the formulation of energy policy for Europe and its potential as a future energy supplier, while expressing real concern at the ongoing race for natural resources and the resulting risks in terms of security and sustainable development.[176] The Commission Communication of 20 November 2008, entitled *The European Union and the Arctic region*, is part of the series of targeted actions launched under the Integrated Maritime Policy Action Plan; it aims to lay the foundations for further reflection on the role of the Union in the Arctic and to achieve a structured and coordinated approach based on the sustainable use of resources, which naturally includes hydrocarbons.[177] The Commission insists that the region's resources could contribute to increasing the EU's security of supply, but that their exploitation should be carried out in full compliance with strict environmental criteria, taking into account the particular vulnerability of the region; it presses in particular for "the introduction of binding international standards."[178]

As for the relevant provisions of the Directive on safety of offshore oil and gas operations of 12 June 2013,[179] they naturally remain very consensual,

---

European Parliament and the Council, An integrated European Union policy for the Arctic, Brussels, 27.4.2016, JOIN(2016) 21 final <https://eur-lex.europa.eu/legal-content/EN/TXT/PDF/?uri=CELEX:52016JC0021&from=EN> (accessed 10 October 2020).

175 Nengye Liu, 'The European Union's Potential Contribution to Enhanced Governance of Offshore Oil and Gas Operations in the Arctic' (2015) *Review of European, Comparative and International Environmental Law*, 223; Nathalie Ros, 'L'Arctique face au changement climatique' [2013] *Journal du Droit international Clunet* 363, especially 397-400.

176 P6_TA(2008)0474, European Parliament resolution of 9 October 2008 on Arctic governance, § 11-13 <http://www.europarl.europa.eu/sides/getDoc.do?pubRef=-//EP//TEXT+TA+P6-TA-2008-0474+0+DOC+XML+V0//EN> (accessed 10 October 2020).

177 Communication from the Commission to the European Parliament and the Council, The European Union and the Arctic Region, Brussels, 20.11.2008, COM(2008) 763 final <https://eur-lex.europa.eu/legal-content/EN/TXT/PDF/?uri=CELEX:52008DC0763&from=EN> (accessed 10 October 2020).

178 Point 3.1, 6-7.

179 Directive 2013/30/EU of the European Parliament and of the Council of 12 June 2013 on safety of offshore oil and gas operations and amending Directive 2004/35/EC, *Official Journal of the European Union*, 28.6.2013, L 178/66-106

even though the idea of a moratorium on drilling in the Arctic had been actively defended by environmentalists in the European Parliament. Recital 52 of the Preamble expressly calls for the strengthening of the protection of the especially fragile marine environment of this ocean facing the exponential development of offshore activities,[180] and Article 33 § 3 merely states that "the Commission shall promote high safety standards for offshore oil and gas operations at international level in relevant global and regional fora, including those relating to Arctic waters."[181]

However, the Directive obviously does not apply in the Arctic,[182] which is under the jurisdiction of the coastal States[183] and the three other circumpolar States.[184] The legal framework thus remains very fragmentary since it consists solely of non-binding regional acts,[185] such as the Artic

---

<https://eur-lex.europa.eu/legal-content/EN/TXT/PDF/?uri=CELEX:32013L0030&from=EN> accessed October 10, 2020.

180 'The Arctic waters are a neighbouring marine environment of particular importance for the Union, and play an important role in mitigating climate change. The serious environmental concerns relating to the Arctic waters require special attention to ensure the environmental protection of the Arctic in relation to any offshore oil and gas operation, including exploration, taking into account the risk of major accidents and the need for effective response. Member States who are members of the Arctic Council are encouraged to actively promote the highest standards with regard to environmental safety in this vulnerable and unique ecosystem, such as through the creation of international instruments on prevention, preparedness and response to Arctic marine oil pollution, and through building, *inter alia*, on the work of the Task Force established by the Arctic Council and the existing Arctic Council Offshore Oil and Gas Guidelines'; L 178/72, Recital 52.

181 Article 33 *Coordinated approach towards the safety of offshore oil and gas operations at international level*, L 178/87.

182 Cécile Pelaudeix, 'Governance of Arctic Offshore Oil & Gas Activities: Multilevel Governance & Legal Pluralism at Stake' (2015) *Arctic Yearbook* 214; Nathalie Ros, 'L'Arctique face au changement climatique' (2013) *Journal du Droit international Clunet* 363, 388 and seq.

183 Of the five Arctic Ocean coastal States - Canada, Denmark, Norway, Russia, and the United States - only Denmark is a Member of the EU.

184 The eight circumpolar States are Canada, Denmark, Finland, Iceland, Norway, Russia, Sweden, and the United States.

185 The only hard law text deals with another issues; it is the Agreement on Cooperation on Marine Oil Pollution Preparedness and Response in the Arctic, adopted on 15 May 2013 <https://oaarchive.arctic-council.org/bitstream/handle/11374/529/EDOCS-2068-v1-ACMMSE08_KIRUNA_2013_agreement_on_oil_pollution_preparedness_and_response_signedAppendices_Original_130510.PDF?sequence=6&isAllowed=y> (accessed 10 October 2020).

Offshore Oil and Gas Guidelines adopted in 1997[186] within the Arctic Council, and revised in 2002[187] and 2009.[188]

Paradoxically, the European Union faces almost similar problems and difficulties in the Black Sea.

3.2.1.1.2 In the Black Sea

Unlike the Arctic Ocean, the Black Sea is one of the marine regions of the European Union, according to Article 4 of the Marine Strategy Framework Directive of 17 June 2008.[189] Indeed, part of the Black Sea is part of the European Union's marine waters, since Bulgaria and Romania became Members of the EU in 2007.[190] The EU Member States are thus obliged to cooperate, in all the regional seas bordering the EU, with all the States of the region, including the activities carried out in the framework of the regional seas conventions; this therefore applies to the Black Sea Commission, even if it is not directly mentioned as such by Article 3 § 10 of the Framework Directive.[191]

---

186 Fourth Ministerial Conference on the Arctic Environmental Protection, Alta, Norway, June 12-13, 1997 <http://www.pame.is/images/03_Projects/Offshore_Oil_and_Gas/Offshore_Oil_and_Gas/oilandgasguidelines.pdf> (accessed 10 October 2020).

187 Protection of the Arctic Marine Environment, PAME Working Group, Third Arctic Council Ministerial Meeting, Inari, Finland, October 9-10, 2002 <https://oaarchive.arctic-council.org/bitstream/handle/11374/1591/MM03_PAME_Attachment_4.pdf?sequence=5&isAllowed=y> (accessed 10 October 2020).

188 Sixth Arctic Council Ministerial Meeting, Tromsø, Norway, April 29, 2009; Arctic Council, Protection of the Arctic Marine Environment Working Group, Arctic Offshore Oil and Gas Guidelines, April 29, 2009 <https://oaarchive.arctic-council.org/bitstream/handle/11374/63/Arctic-Guidelines-2009-13th-Mar2009.pdf?sequence=1&isAllowed=y> (accessed 10 October 2020).

189 Directive 2008/56/EC of the European Parliament and of the Council of 17 June 2008 establishing a framework for community action in the field of marine environmental policy (Marine Strategy Framework Directive), *Official Journal of the European Union*, 25.6.2008, L 164/25-26 <https://eur-lex.europa.eu/legal-content/EN/TXT/PDF/?uri=CELEX:32008L0056&from=EN> (accessed 10 October 2020).

190 Anne-Sophie Lamblin-Gourdin, 'L'intégration de la mer Noire dans l'Union européenne' (2008) *Annuaire de Droit Maritime et Océanique* 97; Gabriela A. Oanta, 'La Unión europea ribereña de un nuevo mar: el mar Negro' in Mariano Aznar Gómez (Coord.), *Estudios de Derecho Internacional y Derecho Europeo en Homenaje al Profesor Manuel Pérez González* (Tirant lo Blanch 2012) Tomo II 1705.

191 Article 3 Definitions: 'For the purposes of this Directive the following defini-

Nevertheless, the European Commission is actively involved in the work of the Black Sea Commission, the executive body of the 1992 Convention on the Protection of the Black Sea against Pollution, to which the six coastal States are Parties.[192] The Union's accession to this convention is even presented as a priority, in particular in the framework of the 2007 Black Sea Synergy,[193] which complements the EU strategy towards Turkey, the European Neighbourhood Policy ("ENP") and the strategic partnership with Russia.[194] However, more than 13 years later, the Union is still not a Party to the regional system formed by the Bucharest Convention on the Protection of the Black Sea Against Pollution of 21 April 1992[195] and its protocols.[196] Indeed, the 1992 Convention only provides for the participation of States and its text would therefore have to be amended on this

---

tions shall apply: [...] 10. 'regional sea convention' means any of the international conventions or international agreements together with their governing bodies established for the purpose of protecting the marine environment of the marine regions referred to in Article 4, such as the Convention on the Protection of the Marine Environment of the Baltic Sea, the Convention for the Protection of the Marine Environment of the North-east Atlantic and the Convention for the Marine Environment and the Coastal Region of the Mediterranean Sea.'

192   Bulgaria, Georgia, Romania, Russia, Turkey, and Ukraine.

193   Communication from the Commission to the Council and the European Parliament, Black Sea Synergy - A New Regional Cooperation Initiative, Brussels, 11.4.2007, COM(2007) 160 final <https://eur-lex.europa.eu/legal-content/EN/TXT/PDF/?uri=CELEX:52007DC0160&from=en> (accessed 10 October 2020).

194   Gabriela A. Oanta, 'The Marine Strategy of the European Union in the Black Sea in the Light of the Current International Regional Context' in Angela Del Vecchio and Fabrizio Marrella (eds.), *International Law and Maritime Governance. Current Issues and Challenges for the Regional Economic Integration Organizations / Droit international et gouvernance maritime. Enjeux actuels et défis pour les organisations régionales d'intégration économique / Diritto Internazionale e Governance Marittima. Problemi Attuali e Sfide per le Organizzazioni di Integrazione Economica Regionale*, Cahiers de l'Association internationale du Droit de la Mer 3 (Editoriale Scientifica 2016) 243.

195   <http://www.blacksea-commissionorg/_convention-fulltext.asp> (accessed 10 October 2020).

196   There are currently four protocols in force: Protocol on the Protection of the Black Sea Marine Environment Against Pollution from Land-based Sources and Activities; Protocol on Cooperation in combating pollution of the Black Sea Marine Environment by Oil and Other Harmful Substances in Emergency Situations; Protocol on The Protection of The Black Sea Marine Environment Against Pollution by Dumping; Black Sea Biodiversity and Landscape Conservation Protocol.

point, which does not seem reasonably foreseeable in the short term, not of course from the point of view of legal technique but because of the opposition of Russia and Turkey, a fortiori in view of the recent geopolitical developments in this highly strategic region.[197]

Although energy interests, in particular oil and gas, exist in the Black Sea, especially around Serpents' Island attributed to Ukraine by the judgment rendered by the International Court of Justice on 3 February 2009,[198] but in the region in which Romania has also been granted rights,[199] the regional system does not directly address these issues and the challenges related to offshore operations. Indeed, there is no dedicated protocol and the umbrella-convention addresses the issue in very general terms, in Article XI Pollution from activities on the continental shelf.[200] This provision establishes that the Parties "shall, as soon as possible, adopt laws and regulations and take measures to prevent, reduce and control pollution of the marine environment of the Black Sea caused by or connected with activities on its continental shelf" (§ 1) and that they "shall cooperate in this field, as appropriate, and endeavour to har-

---

197 Ioannis Stribis, 'Questions actuelles de droit de la mer en mer Noire' in Nathalie Ros et Florence Galletti (eds.), *Le droit de la mer face aux 'Méditerranées:' Quelle contribution de la Méditerranée et des mers semi-fermées au développement du droit international de la mer?* Cahiers de l'Association internationale du Droit de la Mer 5 (Editoriale Scientifica 2016) 183.

198 *Maritime Delimitation in the Black Sea* (Romania v. Ukraine) [2009] ICJ Reports 61 <https://www.icj-cij.org/files/case-related/132/132-20090203-JUD-01-00-EN.pdf> (accessed 10 October 2020).

199 Gabriela A. Oanta, « La incidencia de la Isla de las Serpientes del Mar Negro en la delimitación de los espacios marinos » in José Manuel Sobrino Heredia (ed.), *La contribution de la Convention des Nations Unies sur le droit de la mer à la bonne gouvernance des mers et des océans / La contribución de la Convención de las Naciones Unidas sobre el Derecho del Mar a la buena gobernanza de los mares y océanos / The Contribution of the United Nations Convention on the Law of the Sea to Good Governance of the Oceans and Seas*, Cahiers de l'Association internationale du Droit de la Mer 2 (Editoriale Scientifica 2014) Volume I 363.

200 Article XI Pollution from activities on the continental shelf: '1. Each Contracting Party shall, as soon as possible, adopt laws and regulations and take measures to prevent, reduce and control pollution of the marine environment of the Black Sea caused by or connected with activities on its continental shelf, including the exploration and exploitation of the natural resources of the continental shelf. The Contracting Parties shall inform each other through the Commission of the laws, regulations and measures adopted by them in this respect. 2. The Contracting Parties shall cooperate in this field, as appropriate, and endeavour to harmonize the measures referred to in paragraph 1 of this Article.'

monize" them (§ 2).²⁰¹ Such a provision cannot of course be sufficient to effectively address the environmental challenge posed by these industrial activities, especially in a semi-enclosed sea, known for its vulnerability, such as the Black Sea, and the European Union has little capacity to contribute to the effective improvement of the legal regulation of offshore activities in this marine region, in particular because it is not a Party to the Regional Sea Convention.

However, this aspect probably needs to be put into perspective, because this kind of difficulties also exists when the UE is a Party to Regional Treaties.

3.2.1.2 When the UE is a Party to Regional Treaties

This situation occurs in two other "shared sea-basins"²⁰² of the European Union,²⁰³ within the framework of the regional systems in the Baltic Sea and in the OSPAR Area, even though the EU is a Party to the corresponding conventions.

3.2.1.2.1 In the Baltic Sea

In the Baltic, the regional system is contemporary of the 1970s, with the first Helsinki Convention of 22 March 1974,²⁰⁴ but it is now founded on

---

201   Ioannis Stribis, 'Questions actuelles de droit de la mer en mer Noire' in Nathalie Ros et Florence Galletti (eds.), *Le droit de la mer face aux 'Méditerranées:' Quelle contribution de la Méditerranée et des mers semi-fermées au développement du droit international de la mer?* Cahiers de l'Association internationale du Droit de la Mer 5 (Editoriale Scientifica 2016) 183, at 209-216.

202   Communication from the Commission to the European Parliament, the Council, the European Economic and Social Committee and the Committee of the Regions, Developing the international dimension of the Integrated Maritime Policy of the European Union, Brussels, 15.10.2009, COM(2009)536 final <http://www.europarl.europa.eu/RegData/docs_autres_institutions/commission_europeenne/com/2009/0536/COM_COM(2009)0536_EN.pdf> (accessed 10 October 2020).

203   Annina Cristina Bürgin, 'The Integrated Maritime Policy and Sea Basin Strategies: A comparative analysis of the Atlantic and Baltic Sea Strategies' in Laura Carballo Piñeiro (Coord.), *Retos presentes y futuros de la Política Marítima Integrada de la Unión Europea* (Editorial Bosch 2017) 387.

204   United Nations Treaty Series, 1988, Vol. 1507, 1-25986, 167 <https://treaties.un.org/doc/Publication/UNTS/Volume%201507/volume-1507-I-25986-English.pdf> (accessed 10 October 2020).

the Convention on the Protection of the Marine Environment of the Baltic Sea Area, known as the Helsinki Convention and revised on 9 April 1992.[205] This Convention was approved by the European Community by Council Decision of 21 February 1994 (94/157/EC),[206] and the Community acceded to the 1974 Convention on the same day.[207] Since 2004, when the Baltic States joined the EU, eight of the nine States bordering the Baltic Sea are Members of the European Union,[208] the only exception being Russia.[209] Union Law therefore enjoys a very wide degree of applicability in the region.

Article 12 of the 1992 Convention deals with Exploration and Exploitation of the Seabed and its Subsoil; it is more detailed than Article 10 of the 1974 Convention[210] and establishes in particular that "each Contracting Party shall take all measures in order to prevent pollution of the marine

---

205 Convention on the Protection of the Marine Environment of the Baltic Sea Area, 1992, *Official Journal of the European Communities*, 16. 3. 94, No L 73/20-45 <https://eur-lex.europa.eu/legal-content/EN/TXT/PDF/?uri=CELEX:21994A0316(02)&from=EN> (accessed 10 October 2020).

206 Council Decision of 21 February 1994 on the conclusion, on behalf of the Community, of the Convention on the Protection of the Marine Environment of the Baltic Sea Area (Helsinki Convention as revised in 1992) (94/157/EC), *Official Journal of the European Communities*, 16. 3. 94, No L 73/19 <https://eur-lex.europa.eu/legal-content/EN/TXT/PDF/?uri=CELEX:31994D0157&from=EN> accessed October 10, 2020.

207 Council Decision of 21 February 1994 on the accession of the Community to the Convention on the Protection of the Marine Environment of the Baltic Sea Area 1974 (Helsinki Convention) 94/156/EC, *Official Journal of the European Communities*, 16. 3. 94, No L 73/1 <https://eur-lex.europa.eu/resource.html?uri=cellar:98cb3037-8eb9-474b-81d3-e481fc896a78.0008.02/DOC_2&format=PDF> (accessed 10 October 2020).

208 Germany, Denmark, Estonia, Finland, Latvia, Lithuania, Poland, and Sweden.

209 Michel Voelckel, 'La mer Baltique, quelle identité?' in Nathalie Ros et Florence Galletti (eds.), *Le droit de la mer face aux 'Méditerranées,' Quelle contribution de la Méditerranée et des mers semi-fermées au développement du droit international de la mer?* Cahiers de l'Association internationale du Droit de la Mer 5 (Editoriale Scientifica 2016) 277. The author evokes what he calls in French the '*communautarisation*,' in English the 'communitarization' of the Baltic.

210 Article 10 Exploration and exploitation of the sea-bed and its subsoil: 'Each Contracting Party shall take all appropriate measures in order to prevent pollution of the marine environment of the Baltic Sea Area resulting from exploration or exploitation of its part of the sea-bed and its subsoil or from any associated activities thereon. It shall also ensure that adequate equipment is at hand to start an immediate abatement of pollution in that area.'

environment of the Baltic Sea area resulting from exploration or exploitation of its part of the sea bed and the subsoil thereof or from any associated activities thereon as well as to ensure that adequate preparedness is maintained for immediate response actions against pollution incidents caused by such activities."(§ 1) Paragraph 2[211] refers to Annex VI Prevention of Pollution from Offshore Activities defined as "any exploration and exploitation of oil and gas by a fixed or floating offshore installation or structure including all associated activities thereon."[212] However, the provisions of the Annex are also very general and flexible; in this spirit, they refer to the Use of best available technology and best environmental practice (Regulation 2), but also to the Environmental impact assessment and monitoring (Regulation 3), while attempting to limit Discharges on the exploration phase (Regulation 4) and exploitation phase (Regulation 5).

But the stakes associated with oil and gas activities are actually very limited in the Baltic Sea, even if prospects are supposed to exist off the German coasts. Neither the HELCOM Commission,[213] the executive body of the Helsinki Convention, nor the EU Strategy for the Baltic Sea Region,[214] whose links with the Integrated Maritime Policy have been highlighted by the Commission, address these issues or take into account the possibility of such activities.

The situation is both different and comparable in the OSPAR Area where the EU is also a member but where, on the contrary, offshore activities are already a reality.

3.2.1.2.2 In the OSPAR Area

In the OSPAR Area, the European Union is also a Party to the regional convention, the Convention for the Protection of the Marine Environ-

---

211 Article 12 § 2: 'In order to prevent and eliminate pollution from such activities the Contracting Parties undertake to implement the procedures and measures set out in Annex VI, as far as they are applicable.'
212 Annex VI, Regulation 1 § 1.
213 <https://helcom.fi/> (accessed 10 October 2020).
214 Communication from the Commission to the European Parliament, the Council, the European Economic and Social Committee and the Committee of the Regions concerning the European Union Strategy for the Baltic Sea Region, Brussels, 10.6.2009, COM(2009) 248 final <http://www.europarl.europa.eu/meetdocs/2009_2014/documents/com/com_com(2009)0248_/com_com(2009)0248_en.pdf> (accessed 10 October 2020).

ment of the North-East Atlantic,[215] known as the Paris Convention of 22 September 1992, according to the Council Decision 98/249/EC of 7 October 1997 on the conclusion of the OSPAR Convention;[216] it takes its name from the two conventions it replaces:[217] the 1972 Oslo Convention (OS) on dumping,[218] the scope of which was extended to land-based pollution *lato sensu*, so that to include pollution resulting from the oil industry, by the 1974 Paris Convention (PAR).[219] In addition, 11 of the 15 States Parties[220] to the OSPAR Convention are also Members of the EU,[221] since the OSPAR Area corresponds to the North-East Atlantic Ocean, including the North Sea and the southern part of the Arctic waters. It is the marine region officially referred to as the "North-East Atlantic Ocean," and defined by Article 4 § 2 (a) of the Marine Strategy Framework Directive as including "(i) the Greater North Sea, including the Kattegat, and the English Channel; (ii) the Celtic Seas; (iii) the Bay of Biscay and the Iberian Coast; (iv) in the Atlantic Ocean, the Macaronesian biogeographic region, being the waters surrounding the Azores, Madeira and the Canary Islands."[222] The OSPAR Area is therefore a re-

---

215 <https://www.ospar.org/site/assets/files/1290/ospar_convention_e_updated_text_in_2007_no_revs.pdf> (accessed 10 October 2020).

216 Council Decision of 7 October 1997 on the conclusion of the Convention for the protection of the marine environment of the north-east Atlantic (98/249/EC), *Official Journal of the European Communities*, 3.4.1998, L 104/1 <https://eur-lex.europa.eu/legal-content/EN/TXT/HTML/?uri=CELEX:31998D0249&from=EN> (accessed 10 October 2020).

217 Alexandre Charles Kiss, 'Récents traités régionaux concernant la pollution de la mer' (1976) *Annuaire français de droit international* 720.

218 Convention for the prevention of marine pollution by dumping from ships and aircraft, Oslo 15 February 1972, United Nations Treaty Series, 1974, N° 13269, 4 <https://treaties.un.org/doc/Publication/UNTS/Volume%20932/volume-932-I-13269-English.pdf> (accessed 10 October 2020).

219 Convention for the prevention of marine pollution from land-based sources, Paris 4 June 1974, United Nations Treaty Series, 1989, Vol. 1546, 1-26842, 119 <https://treaties.un.org/doc/Publication/UNTS/Volume%201546/volume-1546-I-26842-English.pdf> (accessed 10 October 2020).

220 The four other OSPAR members are Iceland, Norway, Switzerland, and the United Kingdom.

221 Belgium, Denmark, Finland, France, Germany, Ireland, Luxembourg, Netherlands, Portugal, Spain, and Sweden.

222 Directive 2008/56/EC of the European Parliament and of the Council of 17 June 2008 establishing a framework for community action in the field of marine environmental policy (Marine Strategy Framework Directive), *Official Journal of*

gion highly integrated into EU waters.[223]

The OSPAR Convention includes a specific provision; its Article 5 is devoted to Pollution from offshore sources and provides that "The Contracting Parties shall take, individually and jointly, all possible steps to prevent and eliminate pollution from offshore sources in accordance with the provisions of the Convention, in particular as provided for in Annex III." Annex III on the prevention and elimination of pollution from offshore sources forms an integral part of the Convention under Article 14. These are very general provisions belonging to material soft law, since their conventional nature goes hand in hand with their very low degree of normativity. They are complemented by non-binding provisions, i.e. falling within the domain of formal soft law, the Objectives, Guiding Principles and Strategic Directions for Offshore Oil and Gas Activities adopted in 2010 under the OSPAR Convention for the Protection of the Marine Environment of the North-East Atlantic (2010-2020).[224]

The flexibility of the resulting legal framework clearly reflects the willingness of the Parties to the OSPAR Convention not to regulate too tightly the activity of this sector of their domestic industry, whereas offshore operations are already very developed in the region and called to be even more so in the future. The European Union and its Member States are very much involved in this approach, which ultimately favors a form of industrial and transnational self-regulation, setting only flexible and voluntary frameworks,[225] rather than seeking to develop genuine interna-

---

the European Union, 25.6.2008, L 164/26 <https://eur-lex.europa.eu/legal-content/EN/TXT/PDF/?uri=CELEX:32008L0056&from=EN> (accessed 10 October 2020).

223 In the Atlantic, the reference document is the Communication from the Commission to the European Parliament, the Council, the European Economic and Social Committee and the Committee of the Regions, Action Plan for a Maritime Strategy in the Atlantic area Delivering smart, sustainable and inclusive growth (Text with EEA relevance), Brussels, 13.5.2013, COM(2013) 279 final <https://eur-lex.europa.eu/legal-content/EN/TXT/PDF/?uri=CELEX:52013DC0279&from=EN> (accessed 10 October 2020).

224 The North-East Atlantic Environment Strategy, Strategy of the OSPAR Commission for the Protection of the Marine Environment of the North-East Atlantic 2010-2020, (OSPAR Agreement 2010-3) 19 <http://www.ospar.org/site/assets/files/1413/10-03e_nea_environment_strategy.pdf> (accessed 10 October 2020).

225 Judith van Leeuwen, 'Who greens the waves? Changing authority in the envi-

tional cooperation.[226]

From this point of view, the situation is obviously different in a Hard Law Framework: the Exception of the Mediterranean being one of the rare examples of a sectoral agreement dedicated to offshore, the only one applicable, even partially, in Europe.[227]

3.2.2 In a Hard Law Framework: the Exception of the Mediterranean

As regards regional seas, as well as enclosed and semi-enclosed seas to which it gave its name, the Mediterranean Sea often serves as a model.[228] The regulation of offshore activities provides an excellent illustration, since the Mediterranean is the only shared sea-basin and even the only European marine region where there is a specific international convention, a Protocol Dedicated to Offshore Activities, the EU being a Party to the Protocol.

3.2.2.1 A Protocol Dedicated to Offshore Activities

Still considered pioneering and innovative, more than 25 years after its adoption, the 1994 Offshore Protocol is also fully integrated in the Framework of the Barcelona System, the first and most developed of the

---

ronmental governance of shipping and offshore oil and gas production' (Wageningen Academic Publishers, 2010) 101-151.

226 Nathalie Ros, 'La coopération en droit international de la mer' in Nicolas Guillet (ed.), *Penser le Maritime* (Presse universitaire de Rouen et du Havre 2021), forthcoming.

227 The second exception is the Protocol Concerning Marine Pollution Resulting from Exploration and Exploitation of the Continental Shelf, adopted in 1989 and in force since 1990, within the framework of the 1978 Kuwait Regional Convention for Cooperation on the Protection of the Marine Environment from Pollution, in the ROPME (Regional Organization for the Protection of the Marine Environment) Sea Area which includes the Persian (or Arabian) Gulf and the Gulf of Oman. A third dedicated convention has been adopted on 2 July 2019, the Malabo Protocol on Environmental Standards and Guidelines for Offshore Oil and Gas Activities, within the framework of the Convention on Cooperation for the Protection, Management and Development of the Marine Environment and Coastal Areas of West, Central and Southern African Region (Abidjan Convention).

228 Nathalie Ros et Florence Galletti (eds.), *Le droit de la mer face aux 'Méditerranées,' Quelle contribution de la Méditerranée et des mers semi-fermées au développement du droit international de la mer?* Cahiers de l'Association internationale du Droit de la Mer 5 (Editoriale Scientifica 2016).

regional seas systems of the United Nations Environment Programme ("UNEP").

3.2.2.1.1 In the Framework of the Barcelona System

The European Community, and today the Union, was from the outset a Party to the Mediterranean regional system, and in particular to its legal dimension,[229] based on the Convention for the Protection of the Mediterranean Sea against Pollution of 16 February 1976,[230] now Convention for the Protection of the Marine Environment and the Coastal Region of the Mediterranean of 10 June 1995,[231] and its seven additional and thematic protocols.[232] The European Union is a very active member of the

---

229 On the legal dimension of the Barcelona System: in an allegorical perspective, Evangelos Raftopoulos, 'Theorizing about Conventional Environmental Sea Regimes as International Trusts: The Case of the Barcelona Convention System' in Rüdiger Wolfrum, Maja Seršić, Trpimir M. Šošić (Eds.), *Contemporary Developments in International Law. Essays in Honor of Budislav Vukas* (Brill Nijhoff 2015) 263; for a pragmatic analysis in relation to Mediterranean governance, Nathalie Ros, 'Régimes juridiques et gouvernance internationale de la mer Méditerranée' in Stéphane Doumbé-Billé et Jean-Marc Thouvenin (eds.), Mélanges en l'honneur du Professeur Habib Slim, Ombres et lumières du droit international (Pedone 2016) 205; Nathalie Ros, 'La gouvernance de la mer Méditerranée' in Bogdan Aurescu, Alain Pellet, Jean-Marc Thouvenin et Ion Gâlea (eds.), *Actualité du droit des mers fermées et semi-fermées* (Pedone 2019) 109; with an evolutionary approach, Tullio Scovazzi, 'The protection of the marine environment in the Mediterranean: ideas behind the updating of the 'Barcelona system'' in Giuseppe Cataldi (ed.), *La Méditerranée et le droit de la mer à l'aube du 21ᵉ siècle / The Mediterranean and the Law of the Sea at the Dawn of the 21ˢᵗ Century* (Bruylant 2002) 269.

230 <http://wedocs.unep.org/bitstreamid/53143/convention_eng.pdf> (accessed 10 October 2020).

231 The 1995 Convention entered into force on 9 July 2004 <https://wedocs.unep.org/bitstream/handle/20.500.11822/7096/Consolidated_BC95_Eng.pdf?sequence=1&isAllowed=y> (accessed 10 October 2020).

232 Protocol for the Prevention of Pollution in the Mediterranean Sea by Dumping from Ships and Aircraft, adopted in 1976 and entered into force in 1978; Protocol Concerning Cooperation in Preventing Pollution from Ships and, in Cases of Emergency, Combating Pollution of the Mediterranean Sea, adopted in 2002 and entered into force in 2004; Protocol for the Protection of the Mediterranean Sea against Pollution from Land-Based Sources and Activities, adopted in 1996 and entered into force in 2008; Protocol Concerning Specially Protected Areas and Biological Diversity in the Mediterranean, adopted in 1995 and entered into force in 1999; Protocol for the Protection of the Mediterranean Sea against Pollution Resulting from Exploration and Exploitation of the Continental Shelf and

system,[233] and eight (Croatia, Cyprus, France, Greece, Italy, Malta, Slovenia, and Spain) of the 21 States Parties to the Barcelona System[234] are also Members States of the EU. The Mediterranean Sea is thus obviously one of the marine regions of the Union according to Article 4 of the Marine Strategy Framework Directive, and paragraph 2 of that provision considers "(b) in the Mediterranean Sea: (i) the Western Mediterranean Sea; (ii) the Adriatic Sea; (iii) the Ionian Sea and the Central Mediterranean Sea; (iv) the Aegean-Levantine Sea."[235] However, the EU Member States are in a minority in the Mediterranean Sea, where Union Law can therefore only partially apply, and they are also in a minority within the EU, which explains why this marine region does not always receive the necessary degree of priority within a Union of 28, or even now of 27. All of these issues are nevertheless at the heart of the Union's concerns, as evidenced in particular by the Communication from the Commission to the Council and the European Parliament Towards an Integrated Maritime Policy for better governance in the Mediterranean, presented in 2009.[236]

Article 7 of the 1995 Convention is specifically dedicated to Pollution resulting from exploration and exploitation of the continental shelf and the seabed and its subsoil; it provides that "The Contracting Parties shall take all appropriate measures to prevent, abate, combat and to the fullest possible extent eliminate pollution of the Mediterranean Sea Area resulting

---

the Seabed and its Subsoil, adopted in 1994 and entered into force in 2011; Protocol on the Prevention of Pollution of the Mediterranean Sea by Transboundary Movements of Hazardous Wastes and their Disposal, adopted in 1996 and entered into force in 2008; Protocol on Integrated Coastal Zone Management in the Mediterranean, adopted in 2008 and entered into force in 2011.

233 The European Union is a Party to the Convention and all its Protocols, with the exception of the 1996 Hazardous Waste Protocol, which it has not signed either.

234 Albania, Algeria, Bosnia and Herzegovina, Cyprus, Croatia, Egypt, France, Greece, Israel, Italy, Lebanon, Libya, Malta, Monaco, Montenegro, Morocco, Slovenia, Spain, Syria, Tunisia, and Turkey.

235 Directive 2008/56/EC of the European Parliament and of the Council of 17 June 2008 establishing a framework for community action in the field of marine environmental policy (Marine Strategy Framework Directive), Official Journal of the European Union, 25.6.2008, L 164/26 <https://eur-lex.europa.eu/legal-content/EN/TXT/PDF/?uri=CELEX:32008L0056&from=EN> (accessed 10 October 2020).

236 Brussels, 11.9.2009, COM(2009) 466 final <https://eur-lex.europa.eu/legal-content/EN/TXT/PDF/?uri=CELEX:52009DC0466&from=EN> (accessed 10 October 2020).

from exploration and exploitation of the continental shelf and the seabed and its subsoil."[237] This is, of course, a very general and flexible provision from a normative point of view, but in accordance with the principle of the umbrella-treaty, this article must be supplemented and detailed by a specific protocol, articulated with the other additional protocols,[238] and in particular with the Protocol Concerning Cooperation in Preventing Pollution from Ships and, in Cases of Emergency, Combating Pollution of the Mediterranean Sea, as well as with the Protocol Concerning Specially Protected Areas and Biological Diversity in the Mediterranean. This specific text is the Protocol for the Protection of the Mediterranean Sea against Pollution Resulting from Exploration and Exploitation of the Continental Shelf and the Seabed and its Subsoil;[239] it is known as the 1994 Offshore Protocol.

3.2.2.1.2 The 1994 Offshore Protocol

The so-called Offshore Protocol is a true conventional act, specially dedicated to the challenges of offshore exploration and exploitation, and it intends to establish a Mediterranean governance of this type of industrial activity.[240] Adopted on 14 October 1994, it was designed in a prospective

---

237 Article 7 of the 1976 Convention did not have a very different wording, although it was naturally significantly less prescriptive: 'The Contracting Parties shall take all appropriate measures to prevent, abate and combat pollution of the Mediterranean Sea Area resulting from exploration and exploitation of the continental shelf and its seabed and its subsoil.'

238 On the Mediterranean system dedicated to offshore pollution, Nathalie Ros, 'Problems of Marine Pollution resulting from Offshore Activities according to International and European Union Law' in Andrea Caligiuri (ed.), *Offshore Oil and Gas Exploration and Exploitation in the Adriatic and Ionian Seas* (Editoriale Scientifica 2015) 34; Nathalie Ros, 'Environmental Challenges of Offshore Activities in International and European Union Law' in Andrea Caligiuri (ed.), *Governance of the Adriatic and Ionian Marine Space*, Cahiers de l'Association internationale du Droit de la Mer 4 (Editoriale Scientifica 2016) 203, especially 203-212.

239 <https://wedocs.unep.org/bitstream/handle/20.500.11822/2961/94ig4_4_protocol_eng.pdf?sequence=1&isAllowed=y> (accessed 10 October 2020).

240 Evangelos Raftopoulos, 'Sustainable Governance of Offshore Oil and Gas Development in the Mediterranean: Revitalizing the Dormant Mediterranean Offshore Protocol' (2010) MEPIELAN E-Bulletin <http://www.mepielan-ebulletin.gr/default.aspx?pid=18&CategoryId=4&ArticleId=29&Article=Sustainable-Governance-of-Offshore-Oil-and-Gas-Development-in-the-Mediterranean:-Revitalizing-the-Dormant-Mediterranean-Offshore-Protocol>

way, as early as 1985,[241] before the development of oil and gas operations, and in order to be fully integrated into the future regional system, in accordance with the environmental principles resulting from the Rio Summit. Aware of the problems specific to semi-enclosed seas, as well as of the extreme vulnerability of the Mediterranean, due to the configuration of the basin but above all to the very low rate of water renewal, the States Parties to the Barcelona System wanted precisely to anticipate future industrial developments and their inevitable environmental consequences.

Perfectly integrated into the conventional system, both from the normative and institutional point of view, the Protocol gives, in an emergency situation, a real operational mission to the Mediterranean Regional Center for Emergency Response against Accidental Marine Pollution ("REMPEC").[242] It adopts a broad and general approach to the exploration and exploitation of the continental shelf, its seabed and its subsoil,[243] and is characterized by its global scope, unlike the Directive adopted by the European Union which is limited to the safety of offshore oil and gas operations.

(accessed 10 October 2020); Nathalie Ros, 'Quel régime juridique pour l'exploitation offshore en Méditerranée?' (2015) *Annuaire de Droit Maritime et Océanique* 205; Nathalie Ros, 'Vers une gouvernance régionale de l'offshore en mer Méditerranée?' in Angela Del Vecchio and Fabrizio Marrella (ed.), *International Law and Maritime Governance. Current Issues and Challenges for the Regional Economic Integration Organizations / Droit international et gouvernance maritime. Enjeux actuels et défis pour les organisations régionales d'intégration économique / Diritto Internazionale e Governance Marittima. Problemi Attuali e Sfide per le Organizzazioni di Integrazione Economica Regionale*, Cahiers de l'Association internationale du Droit de la Mer 3 (Editoriale Scientifica 2016) 219.

241 Indeed, it was at the 4th Ordinary Meeting of the Contracting Parties to the Barcelona Convention (CoP 4), held in Genoa in September 1985, that the preparation of a dedicated protocol was first advocated, pursuant to Article 7 of the Convention.

242 In particular under the terms of Articles 18 *Mutual assistance in cases of emergency* and 26 § 3 *Transboundary pollution*, and even if REMPEC should ultimately play a much more important role in the implementation of the Protocol, as reflected in the dedicated Action Plan.

243 On the specific issues of the exploration and exploitation of the Mediterranean continental shelf, in particular with regard to environmental protection requirements, see Nathalie Ros, 'Exploration, Exploitation and Protection of the Mediterranean Continental Shelf' in Claudia Cinelli and Eva Maria Vásquez Gómez (eds.), *Regional Strategies to Maritime Security: a Comparative Perspective* (Tirant lo Blanch, 2014) 101.

From a spatial vantage point, it applies to all areas of the Mediterranean likely to be affected by the consequences of this type of pollution, i.e., the entire Mediterranean Sea and its coastline.[244] All the mineral resources of the continental shelf are taken into account, "whether solid, liquid or gaseous,"[245] and the Protocol is therefore potentially applicable beyond conventional oil and gas activities, especially if methane hydrates and rare earths are actually present in the Mediterranean subsoil. As regards the activities,[246] the Protocol regulates the whole industrial process of exploration and exploitation, including scientific research and the removal of installations;[247] the concept of installation is also broadly defined so that not only drilling and production, but also storage and transport are taken into account.[248]

Of course, the 94 Protocol inevitably includes provisions of material soft law and does not directly incorporate the latest legal and technological innovations, which are nevertheless generally applicable mutatis mutandis under the umbrella-treaty.[249] However, it appears to be characterized by its high level of requirements as well as by its global normative ambition.[250] The most emblematic provision is the obligation of insurance or

---

244 Article 2 *Geographical Coverage*.

245 Article 1 (c).

246 Article 1 (d).

247 This is also provided for in Article 20 *Removal of installations*.

248 Article 1 (f): '"installation" means any fixed or floating structure, and any integral part thereof, that is engaged in activities, including in particular: (i) Fixed or mobile offshore drilling units; (ii) Fixed or floating production units including dynamically-positioned units; (iii) Offshore storage facilities including ships used for this purpose; (iv) Offshore loading terminals and transport systems for the extracted products, such as submarine pipelines; (v) Apparatus attached to it and equipment for the reloading, processing, storage and disposal of substances removed from the seabed or its subsoil.'

249 This applies in particular to obligations related to public consultation, which are not expressly provided for in the Protocol, but are applicable under Article 15 *Public Information and Participation* of the 1995 Convention.

250 Some examples of ambitious provisions can be cited, such as: written authorization for exploration and exploitation (Articles 4, 5 & 6); sanctions for breaches of conventional obligations (Article 7); use of the best available techniques and standards to minimize the risk of pollution (Article 8); environmental impact assessments (Article 5 § 1 (a), and Annex IV); mutual assistance in cases of emergency (Article 18); removal of installations (Article 20); insurance and other financial security to cover liability (Articles 5 § 1 (i), and 27).

other financial security to cover the operator's liability, established by Articles 5 § 1 (i)[251] and 27 § 2;[252] this avant-garde obligation, which is unique in all Public International Law, explains eventually the delay in the entry into force of the Protocol and is, moreover, generally the provision most criticized by States reluctant to commit themselves. If the Protocol finally entered into force, 17 years after its adoption, on 24 March 2011, it was thanks to the six ratifications of Albania, Cyprus, Libya, Morocco, Syria and Tunisia, and despite the opposition from European Union Member States, reluctant to impose such obligations on their offshore industry.[253] This situation remains, the EU being a Party to the Protocol.

3.2.2.2 The EU being a Party to the Protocol

Given the opposition of its Member States to the Protocol, the Decision of Accession can unquestionably be considered a Change of Strategy of the EU, due to the accident on the Deepwater Horizon platform on 20 April 2010.

3.2.2.2.1 *A Change of Strategy*

Following the reflection process launched in May 2010, the strategic evolution starts from 12 October 2010, according to the Communication from the Commission to the European Parliament and to the Council entitled Facing the challenge of the safety of offshore oil and

---

251 Article 5 Requirements for authorizations: '1. The Contracting Party shall prescribe that any application for authorization or for the renewal of an authorization is subject to the submission of the project by the candidate operator to the competent authority and that any such application must include, in particular, the following: [...] (i) The insurance or other financial security to cover liability as prescribed in Article 27, paragraph 2 (b).'

252 Article 27 Liability and compensation: '2. Pending development of such procedures, each Party: (a) Shall take all measures necessary to ensure that liability for damage caused by activities is imposed on operators, and they shall be required to pay prompt and adequate compensation; (b) Shall take all measures necessary to ensure that operators shall have and maintain insurance cover or other financial security of such type and under such terms as the Contracting Party shall specify in order to ensure compensation for damages caused by the activities covered by this Protocol.'

253 On the Protocol and the consequences of its entry into force, analyzed from a Euro-Mediterranean perspective, Nathalie Ros, 'La réglementation euro-méditerranéenne des activités offshore' (2015) *Diritto del Commercio Internazionale* 93.

gas activities.[254] Indeed, the Commission had proposed to the Council to sign the Protocol as early as 1992,[255] i.e., before its adoption, but the text had so far been neither signed nor ratified, and the accident that occurred on the Deepwater Horizon platform in April 2010 is therefore undoubtedly the real cause of the change in the European Union's attitude. In its Communication, the Commission recommends relaunching the process, in close collaboration with the Member States concerned, in order to allow the entry into force of the Mediterranean Offshore Protocol. This proposal is supported by the European Parliament in its Resolution of 13 September 2011 on facing the challenge of the safety of offshore oil and gas activities,[256] while the Protocol has in the meantime entered into force in March ... On 27 October 2011,[257] the Commission publishes a Proposal for a Council Decision on the accession of the European Union to the Protocol for the Protection of the Mediterranean Sea against Pollution Resulting from Exploration and Exploitation of the

---

254 Brussels, 12.10.2010, COM(2010) 560 final <https://eur-lex.europa.eu/legal-content/EN/TXT/PDF/?uri=CELEX:52010DC0560&from=EN> (accessed 10 October 2020).

255 The Commission had then adopted, and transmitted to the Council, a proposal for a Council Decision concerning the signature of the future Offshore Protocol to the Barcelona Convention, but at that time it had been considered more appropriate to continue working on a Community system of environmental liability than to anticipate by adopting an international agreement; Commission of the European Communities, Proposal for a Council Decision concerning the Signature of a Protocol for the Protection of the Mediterranean Sea against Pollution resulting from Exploration and Exploitation of the Continental Shelf and the Seabed and its Subsoil, Brussels, 22.09.94, COM(94) 397 final <http://eur-lex.europa.eu/LexUriServ/LexUriServ.do?uri=COM:1994:0397:FIN:EN:PDF> (accessed 10 October 2020).

256 P7_TA(2011)0366, European Parliament resolution of 13 September 2011 on facing the challenges of the safety of offshore oil and gas activities (2011/2072(INI)) <http://www.europarl.europa.eu/sides/getDoc.do?pubRef=-//EP//TEXT+TA+P7-TA-2011-0366+0+DOC+XML+V0//EN> (accessed 10 October 2020). Point 83 'Stresses the importance of bringing fully into force the un-ratified 1994 Mediterranean Offshore Protocol, targeting protection against pollution resulting from exploration and exploitation.'

257 On the same day as the Proposal for a Regulation of the European Parliament and of the Council on safety of offshore oil and gas prospection, exploration and production activities, Brussels, 27.10.2011, COM(2011) 688 final 2011/0309 (COD) <http://www.europarl.europa.eu/meetdocs/2009_2014/documents/com/com_com(2011)0688_/com_com(2011)0688_en.pdf> (accessed 10 October 2020).

Continental Shelf and the Seabed and its Subsoil,[258] which ratifies the ongoing strategic development; the risks induced by offshore operations are prioritized in a global legal context, integrating both the accession to the Protocol and the obligations of the Member States with regard to the Barcelona System, as well as the Marine Strategy Framework Directive and the development of EU Law in this area. It is in this context that the European Parliament finally gives its approval to the EU's accession to the Offshore Protocol on 20 November 2012;[259] the Decision of Accession comes one month later.

3.2.2.2.2 The Decision of Accession

The accession of the EU to the Offshore Protocol is in fact formalized by the Council Decision of 17 December 2012, Article 1 of which provides that "The accession of the European Union to the Protocol for the Protection of the Mediterranean Sea against pollution resulting from exploration and exploitation of the continental shelf and the seabed and its subsoil is hereby approved on behalf of the Union."[260] The Decision explicitly refers to the Mediterranean context from which it appears in fact inseparable, not only with regard to the recent discoveries and promises

---

[258] Proposal for a Council Decision on the accession of the European Union to the Protocol for the Protection of the Mediterranean Sea against pollution resulting from exploration and exploitation of the continental shelf and the seabed and its subsoil, Brussels, 27.10.2011, COM(2011) 690 final, 2011/0304 (NLE) <http://eur-lex.europa.eu/legal-content/EN/TXT/PDF/?uri=CELEX:52011PC0690&from=EN> (accessed 10 October 2020).

[259] P7_TA(2012)0415, EU accession to the Protocol for the Protection of the Mediterranean Sea against pollution resulting from exploration and exploitation of the continental shelf and the seabed and its subsoil, European Parliament legislative resolution of 20 November 2012 on the draft Council Decision on the accession of the European Union to the Protocol for the Protection of the Mediterranean Sea against pollution resulting from exploration and exploitation of the continental shelf and the seabed and its subsoil (09671/2012 – C7-0144/2012 – 2011/0304(NLE)) <http://www.europarl.europa.eu/sides/getDoc.do?type=TA&reference=P7-TA-2012-0415&language=EN> (accessed 10 October 2020).

[260] Council Decision of 17 December 2012 on the accession of the European Union to the Protocol for the Protection of the Mediterranean Sea against pollution resulting from exploration and exploitation of the continental shelf and the seabed and its subsoil (2013/5/EU), *Official Journal of the European Union*, 9.1.2013, L 4/13-14 <http://eur-lex.europa.eu/LexUriServ/LexUriServ.do?uri=OJ:L:2013:004:0013:0014:EN:PDF> (accessed 10 October 2020).

of hydrocarbons and other mineral resources, in particular rare earths, but also to the risks and dangers revealed by the Deepwater Horizon accident, which are multiplied and naturally amplified, to the level of a fatal disaster, in a semi-enclosed and fragile sea such as the Mediterranean. Referring to Directive 2008/56/EC of the European Parliament and of the Council of 17 June 2008 establishing a framework for community action in the field of marine environmental policy, the Marine Strategy Framework Directive,[261] which constitutes the environmental pillar of the Integrated Maritime Policy, the Decision highlights the challenges of a future exploitation of the Mediterranean continental shelf.

Accession to the Protocol therefore appears to mark an undeniable stage in the process of awareness and involvement of the European Union and its Member States in the management of the challenges and risks associated with offshore;[262] even geographically limited to the Mediterranean basin, it must therefore also be understood in conjunction with other normative developments and concomitant progress in EU Law, leading to Integration into European Union Law.

3.3 To Integration into European Union Law

As regards offshore energy activities regulation, the strategy of the EU is obviously twofold, with an internal dimension dedicated to the development of a normative framework, from the European Union Law Acquis to the more specific Contribution of Directive 2013/30/EU.

3.3.1 *The European Union Law Acquis*

As a common basis of rights and obligations for the Members of the Union, the acquis refers of course to the set of legal rules applicable within the EU legal order; but in the case of offshore operations there is a specific acquis, which has to be naturally understood by reference to the substantive law in force for EU Member States, primarily as a result of the Integration of the Mediterranean Acquis.

---

261 https://eur-lex.europa.eu/LexUriServ/LexUriServ.do?uri=OJ:L:2008: 164:0019:0040:EN:PDF> (accessed 10 October 2020).

262 Marie Bourrel, 'L'Union européenne adhère au protocole sur les activités offshore en Méditerranée' (2013) *Droit de l'environnement* 212.

3.3.1.1 Integration of the Mediterranean Acquis

The obligations resulting from Mediterranean International Law are indeed an integral part of the acquis, insofar as they are defined and integrated into European Union Law and for Mediterranean EU Member States.

3.3.1.1.1 Into European Union Law

Indeed, accession automatically implies the integration of the 1994 Offshore Protocol into the legal order of the European Union. The legal framework established by Article 216 of the Treaty on the Functioning of the European Union applies to the Protocol for the Protection of the Mediterranean Sea against Pollution Resulting from Exploration and Exploitation of the Continental Shelf and the Seabed and its Subsoil; this provision provides that "The Union may conclude an agreement with one or more third countries or international organisations where the Treaties so provide or where the conclusion of an agreement is necessary in order to achieve, within the framework of the Union's policies, one of the objectives referred to in the Treaties, or is provided for in a legally binding Union act or is likely to affect common rules or alter their scope" (§ 1), and that "Agreements concluded by the Union are binding upon the institutions of the Union and on its Member States" (§ 2).

The legal consequences of the integration are therefore twofold. The obligations relating to the implementation and application of the Offshore Protocol are not only incumbent on the European Union, but they are also binding for Mediterranean EU Member States, including in the case of almost all of them which have not yet acceded to the Protocol.

3.3.1.1.2 For Mediterranean EU Member States

Indeed, for the eight Member States of the Union that are also Mediterranean coastal States, there is a legal obligation, under EU Law, to transpose the provisions and requirements of the Protocol into their domestic legislation. But the accession of the European Union to the Protocol also entails other consequences and other types of obligations for them, more indirect, this time as Parties to the Mediterranean system. This dual legal dimension is a very large part of the challenges which the system has to face. The integration of the Protocol into EU Law does not exempt the

Mediterranean Member States from ratifying the text, especially as promises deposits exist, and offshore exploration and exploitation activities are already underway, or under development, in waters under the sovereignty or jurisdiction of several of them; their ratification would therefore be neither unnecessary nor superfluous.[263]

In theory, and in general, accession and the obligation to transpose should encourage, and even incite, the EU Mediterranean States to ratify the Protocol, but in practice they continue to be its most fervent opponents. However, the Accession Decision stated at the end of 2012 that "in addition to Cyprus, some other Member States that are Contracting Parties to the Barcelona Convention have announced recently their intention to also ratify the Protocol,"[264] and several of them said they were on the verge of ratification.[265] But eight years later, the status quo remains ... Cyprus, a Party since 2006,[266] and Croatia, a Party since 2018,[267] are still the only Member

---

263 If the Offshore Protocol naturally needs more ratifications, a better balance between EU and non-EU Member States is indeed also necessary in order for the Protocol to become more effective in the whole Mediterranean basin, and enhance cooperation and environmental protection, including developing *Mutual assistance in cases of emergency* (Article 18), providing *Scientific and technical assistance to developing countries* (Article 24), and sharing *Mutual information* (Article 25), to fight against *Transboundary pollution* (Article 26), as required by the Protocol.

264 Council Decision of 17 December 2012 on the accession of the European Union to the Protocol for the Protection of the Mediterranean Sea against pollution resulting from exploration and exploitation of the continental shelf and the seabed and its subsoil (2013/5/EU), Official Journal of the European Union, 9.1.2013, L 4/13, Recital 3 <http://eur-lex.europa.eu/LexUriServ/LexUriServ.do?uri=OJ:L:2013:004:0013:0014:EN:PDF> (accessed 10 October 2020).

265 According to a table integrated into the Final Report of the 3rd Meeting of the Working Group on the Offshore Protocol, and based on the information provided by the States, France which has still not signed the text, Italy and Malta, claimed in 2014 to have initiated internal consultation procedures with a view to ratification; without considering ratification, Greece declared to implement the provisions of the Protocol and Slovenia to discuss its transposition into domestic law; Croatia and Spain had not provided any information on their national situation at that time. Report of the 3rd Offshore Protocol Working Group Meeting, Attard, Malta 17-18 June 2014, REMPEC/WG/35/6, 31 July 2014, 4, § 23, Table 1 *Offshore Protocol ratification process* <https://www.rempec.org/en/knowledge-centre/online-catalogue/final-report-e-consolidated-31-7-14.pdf> (accessed 10 October 2020).

266 Cyprus, a Member State of the European Union since 1 January 2004, ratified the Offshore Protocol on 16 May 2006.

267 Croatia, a Member State of the European Union since 1 July 2013, ratified the

States of the European Union among the Contracting Parties; none of the other Mediterranean EU Member States has acceded to the Protocol since the Union itself became a Party to it. Despite the political promises, the legal situation therefore remains unchanged: Greece, Italy, Malta, Slovenia, and Spain have signed the Protocol but without ratifying it; France has neither signed nor ratified it.[268] For EU Member States, this type of attitude constitutes a real challenge to the EU's legal order, to the acquis which incorporates a large part of the new requirements and even specifies certain modalities,[269] and in fine to the Integrated Maritime Policy.

3.3.1.2 For EU Member States

Member States must, of course, respect the acquis of EU Law, and first of all in this specific area the acquis of International Conventions Integrated into EU Law and Texts of the European Legal Order.

3.3.1.2.1 *International Conventions Integrated into EU Law*

Three international agreements can prima facie be considered particularly relevant to the exploration and exploitation of the continental shelf and its resources.

First of all, of course, is the United Nations Convention on the Law of the Sea of 10 December 1982. Described as a "Constitution for the Oceans,"[270] UNCLOS devotes its Part VI to the Continental Shelf, which defines it as well as its legal regime articulated between rights and obliga-

---

Offshore Protocol on 8 February 2018, with entry into force on 10 March 2018.

268 Outside the European Union, the situation is not much more encouraging. Israel and Monaco have signed the text, but Algeria, Bosnia and Herzegovina, Egypt, Lebanon, Montenegro, and Turkey have neither signed nor ratified.

269 For a detailed and in-depth analysis of the *acquis* in relation to the Offshore Protocol, see Final Report *Safety of offshore exploration and exploitation activities in the Mediterranean: creating synergies between the forthcoming EU Regulation and the Protocol to the Barcelona Convention*, Annex IV: EU *Acquis* (relevant to Offshore Protocol obligations) <http://ec.europa.eu/environment/marine/international-cooperation/regional-sea-conventions/barcelona-convention/pdf/Annex_IV.pdf> (accessed 10 October 2020).

270 'A Constitution for the Oceans,' Remarks by Tommy T. B. Koh, of Singapore, President of the Third United Nations Conference on the Law of the Sea <https://www.un.org/depts/los/convention_agreements/texts/koh_english.pdf> accessed October 10, 2020.

tions of the coastal State.²⁷¹ The coastal State has sovereign rights over the continental shelf and its resources, but these rights can only be balanced by the respect of certain obligations. Part VI should therefore be read in conjunction with Part XII Protection and preservation of the marine environment,²⁷² in particular Articles 208 Pollution from seabed activities subject to national jurisdiction²⁷³ and 214 Enforcement with respect to pollution from seabed activities.²⁷⁴ In the absence of any other universal

---

271 Vicente Marotta Rangel, 'Le plateau continental dans la Convention de 1982 sur le droit de la mer' (1985) *Recueil des cours de l'Académie de droit international de La Haye* 269; Jean-François Pulvenis, 'Le plateau continental. Définition et régime des ressources' in René-Jean Dupuy et Daniel Vignes (eds.), *Traité du nouveau droit de la mer* (Economica Bruylant 1985) 275; Nathalie Ros, 'L'Etat côtier et son plateau continental: enjeux et perspectives dans le nouveau droit de la mer' in *Liber Amicorum Haritini Dipla* (Pedone 2020) 109.

272 Article 192 lays down the *General obligation* under which 'States have the obligation to protect and preserve the marine environment,' and Article 193 Sovereign right of States to exploit their natural resources adds that 'States have the sovereign right to exploit their natural resources pursuant to their environmental policies and in accordance with their duty to protect and preserve the marine environment.' On Part XII (Articles 192 to 237) and its contribution, Alan E. Boyle, 'Marine Pollution under the Law of the Sea Convention' (1985) *American Journal of International Law* 347; Pierre-Marie Dupuy et Martine Rémond-Gouilloud, 'La préservation du milieu marin,' in René-Jean Dupuy et Daniel Vignes (eds.), *Traité du Nouveau Droit de la Mer* (Economica Bruylant 1985) 979; Moira L. McConnell and Edgar Gold, 'The Modern Law of the Sea: Framework for the Protection and Preservation of the Marine Environment' (1991) *Case Western Reserve Journal of International Law* 83.

273 Article 208 *Pollution from seabed activities subject to national jurisdiction*: '1. Coastal States shall adopt laws and regulations to prevent, reduce and control pollution of the marine environment arising from or in connection with seabed activities subject to their jurisdiction and from artificial islands, installations and structures under their jurisdiction, pursuant to articles 60 and 80. 2. States shall take other measures as may be necessary to prevent, reduce and control such pollution. 3. Such laws, regulations and measures shall be no less effective than international rules, standards and recommended practices and procedures. 4. States shall endeavour to harmonize their policies in this connection at the appropriate regional level. 5. States, acting especially through competent international organizations or diplomatic conference, shall establish global and regional rules, standards and recommended practices and procedures to prevent, reduce and control pollution of the marine environment referred to in paragraph 1. Such rules, standards and recommended practices and procedures shall be re-examined from time to time as necessary.'

274 Article 214 *Enforcement with respect to pollution from seabed activities*: 'States shall enforce their laws and regulations adopted in accordance with article 208 and shall adopt laws and regulations and take other measures necessary to imple-

agreement applicable to this type of activity,[275] and given that only three dedicated regional conventions exist to date,[276] the provisions of the Montego Bay Convention appear essential even if they remain very general. The legal regulation of these activities is actually very weak, under Public International Law,[277] due to the reluctance of States to adopt a real sustainable development approach, imposing a minimum of obligations on the oil industry and able to resist the powerful influence of lobbies.[278]

---

    ment applicable international rules and standards established through competent international organizations or diplomatic conference to prevent, reduce and control pollution of the marine environment arising from or in connection with seabed activities subject to their jurisdiction and from artificial islands, installations and structures under their jurisdiction, pursuant to articles 60 and 80.'

275 There is no other universal convention, except in a completely different field, and generally more focused on pollution by ships than by offshore, the IMO Convention on Oil Pollution Preparedness, Response and Cooperation (OPRC). For an analysis more specifically focused on the issues of prevention, preparedness and response to offshore pollution, Sergei Vinogradov, 'The Impact of the Deepwater Horizon: The Evolving International Legal Regime for Offshore Accidental Pollution Prevention, Preparedness, and Response' (2013) *Ocean Development & International Law* 335.

276 In addition to the 1994 Mediterranean Protocol, the Protocol Concerning Marine Pollution Resulting from Exploration and Exploitation of the Continental Shelf, adopted in 1989 and in force since 1990, within the framework of the 1978 Kuwait Regional Convention for Cooperation on the Protection of the Marine Environment from Pollution, in the ROPME (Regional Organization for the Protection of the Marine Environment) Sea Area which includes the Persian (or Arabian) Gulf and the Gulf of Oman, and the third dedicated convention, adopted on 2 July 2019, the Malabo Protocol on Environmental Standards and Guidelines for Offshore Oil and Gas Activities, within the framework of the Convention on Cooperation for the Protection, Management and Development of the Marine Environment and Coastal Areas of West, Central and Southern African Region (Abidjan Convention).

277 John Warren Kindt, 'The Law of the Sea: Offshore Installations and Marine Pollution' (1985) *Pepperdine Law Review* 381; Nathalie Ros, 'La pollution résultant de l'exploitation du sol et du sous-sol: le cas du plateau continental' in *Droit des sites et sols pollués. Bilans et perspectives* (L'Harmattan 2018) 39; for a regional example, Nathalie Ros, 'Exploration, Exploitation and Protection of the Mediterranean Continental Shelf' in Claudia Cinelli and Eva Maria Vásquez Gómez (ed.), *Regional Strategies to Maritime Security: a Comparative Perspective* (Tirant lo Blanch 2014) 101; and in a broader perspective, including the Area, Tullio Treves, 'La pollution résultant de l'exploration et de l'exploitation des fonds marins en droit international' (1978) *Annuaire français de droit international* 827.

278 As an example, it is interesting to recall that France is undoubtedly one of the only States in the world to have waived the levying of all financial and fiscal

The other relevant conventions under the acquis are more directly related to the environmental legal order; both have not only a sectoral but also a regional scope, being two agreements adopted under the auspices of the United Nations Economic Commission for Europe: the Espoo Convention on Environmental Impact Assessment in a Transboundary Context of 25 February 1991;[279] and the Aarhus Convention on Access to Information, Public Participation in Decision-making and Access to Justice in Environmental Matters of 25 June 1998.[280]

However, the principle of the acquis also implies respecting directly or indirectly relevant Texts of the European Legal Order.

3.3.1.2.2 Texts of the European Legal Order

In the case of offshore oil and gas operations, this is a relatively large corpus juris a fortiori in relation to Directive 2013/30/EU. It already includes about 20 European legal acts, directives, regulations and decisions, adopted in four sectoral areas: environment,[281] including marine issues, and

---

rights in the event of offshore exploitation, since 1993 after the Mining Code was discreetly amended to this effect by an amendment to the Finance Act for 1994, never challenged since then by any of the successive governments in power; for a more detailed analysis of French legislation, Nathalie Ros, 'Au-delà de la borne 602 : la frontière maritime entre l'Espagne et la France en mer Méditerranée' (2014) *Journal du Droit international Clunet* 1107-1110; Nathalie Ros, 'La pollution résultant de l'exploitation du sol et du sous-sol: le cas du plateau continental' in *Droit des sites et sols pollués. Bilans et perspectives* (L'Harmattan 2018) 42-43.

279 <https://www.unece.org/fileadmin/DAM/env/eia/documents/legaltexts/Espoo_Convention_authentic_ENG.pdf> accessed October 10, 2020; Wick Schrage, 'The Convention on Environmental Impact Assessment in a Transboundary Context' (1997) *Environmental Law Network International* 21.

280 <https://www.unece.org/fileadmin/DAM/env/pp/documents/cep43e.pdf> (accessed 10 October 2020); Michael Mason, 'Information Disclosure and Environmental Rights: The Aarhus Convention' (2010) *Global Environmental Politics* 10; and also (1999) *Revue Juridique de l'Environnement, La Convention d'Aarhus*, numéro spécial.

281 Council Directive 92/43/EEC of 21 May 1992 on the conservation of natural habitats and of wild fauna and flora; Directive 2001/42/EC of the European Parliament and of the Council of 27 June 2001 on the assessment of the effects of certain plans and programmes on the environment; Directive 2003/4/EC of the European Parliament and of the Council of 28 January 2003 on public access to environmental information and repealing Council Directive 90/313/EEC; Directive 2003/35/EC of the European Parliament and of the Council of 26 May 2003 providing for public participation in respect of the drawing up of

of course Directive 2008/56/EC of the European Parliament and of the Council of 17 June 2008 establishing a framework for community action in the field of marine environmental policy (Marine Strategy Framework Directive)[282] and environmental pillar of Integrated Maritime Policy;[283] safety and health of workers and civil protection;[284] industrial safety;[285] oil

---

certain plans and programmes relating to the environment and amending with regard to public participation and access to justice Council Directives 85/337/EEC and 96/61/EC; Directive 2004/35/EC of the European Parliament and of the Council of 21 April 2004 on environmental liability with regard to the prevention and remedying of environmental damage; Directive 2009/147/EC of the European Parliament and of the Council of 30 November 2009 on the conservation of wild birds; Directive 2011/92/EU of the European Parliament and of the Council of 13 December 2011 on the assessment of the effects of certain public and private projects on the environment; Directive 2014/52/EU of the European Parliament and of the Council of 16 April 2014 amending Directive 2011/92/EU on the assessment of the effects of certain public and private projects on the environment.

282 <https://eur-lex.europa.eu/legal-content/EN/TXT/PDF/?uri=CELEX:32008L0056&from=EN> (accessed 10 October 2020).

283 Yves Auffret, 'La directive stratégie pour le milieu marin : contenu et portée dans le contexte de la mise en œuvre de la politique maritime de l'Union européenne' [2009] *Revue Européenne de Droit de l'Environnement* 171; Ronán Long, 'The Marine Strategy Framework Directive: A new European approach to the regulation of the marine environment, marine natural resources and marine ecological services' [2011] *Journal of Energy and Natural Resources Law* 1.

284 Council Directive of 12 June 1989 on the introduction of measures to encourage improvements in the safety and health of workers at work (89/391/EEC); Council Directive 92/91/EEC of 3 November 1992 concerning the minimum requirements for improving the safety and health protection of workers in the mineral-extracting industry through drilling; Council Decision of 8 November 2007 establishing a Community Civil Protection Mechanism; Directive 2012/18/EU of the European Parliament and of the Council of 4 July 2012 on the control of major-accident hazards involving dangerous substances, amending and subsequently repealing Council Directive 96/82/EC.

285 Directive 94/9/EC of the European Parliament and the Council of 23 March 1994 on the approximation of the laws of the Member States concerning equipment and protective systems intended for use in potentially explosive atmospheres; Council Directive 96/82/EC of 9 December 1996 on the control of major-accident hazards involving dangerous substances; Directive 97/23/EC of the European Parliament and of the Council of 29 May 1997 on the approximation of the laws of the Member States concerning pressure equipment; Directive 2006/42/EC of the European Parliament and of the Council of 17 May 2006 on machinery, and amending Directive 95/16/EC; Regulation (EC) No 1907/2006 of the European Parliament and of the Council of 18 December 2006 concerning the Registration, Evaluation, Authorisation and Restriction of

industry.[286]

With a different approach, and according to a soft organizational logic, the Commission Decision of 19 January 2012 created the European Union Offshore Oil and Gas Authorities Group (EUOAG);[287] it is a group of experts set up with the mission of promoting effective collaboration in this area between the Commission and national representatives, in particular by disseminating best practices and operational intelligence, by setting priorities for strengthening standards, and advising the Commission on any regulatory reform.

Still from an institutional point of view, Regulation (EU) No 100/2013 of the European Parliament and of the Council of 15 January 2013 amending Regulation (EC) No 1406/2002 establishing a European Maritime Safety Agency[288] has also to be integrated into the relevant acquis. This text gives a more important role to the European Maritime Safety Agency ("EMSA"), whose competences are doubly enlarged. From a material vantage point, the 2013 Regulation is precisely in line with the new context of EU action in the field of offshore operations; whereas EMSA was initially created to work on maritime safety and prevention of pollution by ships,[289] the text thus provides a new mission of the Agency in the field

---

Chemicals (REACH); Directive 2008/98/EC of the European Parliament and of the Council of 19 November 2008 on waste and repealing certain Directives; Regulation (EC) No 1272/2008 of the European Parliament and of the Council of 16 December 2008 on classification, labelling and packaging of substances and mixtures, amending and repealing Directives 67/548/EEC and 1999/45/EC, and amending Regulation (EC) No 1907/2006; Directive 2010/75/EU of the European Parliament and of the Council of 24 November 2010 on industrial emissions (integrated pollution prevention and control).

286 Directive 94/22/EC of the European Parliament and the Council of 30 May 1994 on the conditions for granting and using authorizations for the prospection, exploration and production of hydrocarbons.

287 Commission Decision of 19 January 2012 on setting up of the European Union Offshore Oil and Gas Authorities Group (2012/C 18/07), Official Journal of the European Union, 21.1.2012, C 18/8-10 <https://eur-lex.europa.eu/legal-content/EN/TXT/PDF/?uri=CELEX:32012D0121(01)&qid=1511894391710&from=EN> accessed October 10, 2020.

288 OJEU 9.2.2013, L 39/30-40 <https://eur-lex.europa.eu/legal-content/EN/TXT/PDF/?uri=CELEX:32013R0100&from=EN> (accessed 10 October 2020).

289 Regulation (EC) No 1406/2002 of the European Parliament and of the Council of 27 June 2002 establishing a European Maritime Safety Agency, 5.8.2002, L 208/1-9 <https://eur-lex.europa.eu/legal-content/EN/TXT/PD

of response to marine pollution caused by oil and gas installations. However, the Agency's field of competence has also been extended from a geographical point of view, since its services can now no longer benefit only the Member States of the Union,[290] but also "States applying for accession to the Union, and, where applicable, to European Neighbourhood partner countries and to countries taking part in the Paris MoU."[291]

But these achievements[292] can only be understood and analyzed in reference to the adoption of an act dedicated to offshore operations within the Union's legal order, i.e., the Contribution of Directive 2013/30/EU of 12 June 2013.[293]

3.3.2 The Contribution of Directive 2013/30/EU

The Directive on Safety of Offshore Oil and Gas Operations is at the heart of the new approach developed by the European Union following

---

F/?uri=CELEX:32002R1406&from=EN> (accessed 10 October 2020).

290 From a strictly Mediterranean perspective, this evolution explains why EMSA therefore appears potentially able to become involved in the implementation of certain aspects of the Offshore Protocol, in collaboration with REMPEC; on this eventuality, see Final Report *Safety of offshore exploration and exploitation activities in the Mediterranean: creating synergies between the forthcoming EU Regulation and the Protocol to the Barcelona Convention* 64-89 <http://ec.europa.eu/environment/marine/international-cooperation/regional-sea-conventions/barcelona-convention/pdf/Final%20Report%20Offshore%20Safety%20Barcelona%20Protocol%20.pdf> (accessed 10 October 2020).

291 Regulation (EU) No 100/2013 of the European Parliament and of the Council of 15 January 2013 amending Regulation (EC) No 1406/2002 establishing a European Maritime Safety Agency, Official Journal of the European Union, 9.2.2013, L 39/36, Article 2 § 5 <https://eur-lex.europa.eu/legal-content/EN/TXT/PDF/?uri=CELEX:32013R0100&from=EN> (accessed 10 October 2020).

292 See also, P8_TA(2016)0478, Liability, compensation and financial security for offshore oil and gas operations, European Parliament resolution of 1 December 2016 on liability, compensation and financial security for offshore oil and gas operations (2015/2352(INI)) (2018/C 224/25), Thursday 1 December 2016, Official Journal of the European Union, 27.6.2018, C 224/157-162 <https://eur-lex.europa.eu/legal-content/EN/TXT/PDF/?uri=CELEX:52016IP0478&from=EN> (accessed 10 October 2020).

293 Directive 2013/30/EU of the European Parliament and of the Council of 12 June 2013 on safety of offshore oil and gas operations and amending Directive 2004/35/EC, Official Journal of the European Union, 28.6.2013, L 178/66-106 <https://eur-lex.europa.eu/legal-content/EN/TXT/PDF/?uri=CELEX:32013L0030&from=EN> (accessed 10 October 2020).

its awareness raising in 2010; however, its contribution actually appears rather disappointing, and constitutes as many Challenges for EU Blue Growth.

3.3.2.1 The Directive on Safety of Offshore Oil and Gas Operations

In order to be able to objectively assess the Directive, it is necessary to evaluate its Formal Framework: A Directive, not a Regulation, and its Material Framework: Safety of Offshore Oil and Gas Operations.

3.3.2.1.1 Formal Framework: A Directive, not a Regulation

The initiative comes in the wake of the Gulf of Mexico disaster, and therefore also coincides with the procedure for EU accession to the 1994 Protocol.[294] In both cases, it is the Communication from the Commission to the European Parliament Facing the challenge of the safety of offshore oil and gas activities that appears to be at the basis of the reflection process, on 12 October 2010,[295] in connection with the European Parliament Resolution of 7 October 2010 on the EU action on oil exploration and extraction in Europe.[296]

The Commission formally proposes the adoption of a corpus juris specifically dedicated to offshore oil and gas operations and installations. Given the considerable disparities between the laws and practices of the Member States and the gaps in European and international legislation, the Commission's objective is to reduce the risk of a major accident in

---

294 Final Report *Safety of offshore exploration and exploitation activities in the Mediterranean: creating synergies between the forthcoming EU Regulation and the Protocol to the Barcelona Convention* <http://ec.europa.eu/environment/marine/international-cooperation/regional-sea-conventions/barcelona-convention/pdf/Final%20Report%20Offshore%20Safety%20Barcelona%20Protocol%20.pdf> (accessed 10 October 2020).

295 Communication from the Commission to the European Parliament and the Council, Facing the challenge of the safety of offshore oil and gas activities, Brussels, 12.10.2010, COM(2010) 560 final 3 <http://eur-lex.europa.eu/LexUriServ/LexUriServ.do?uri=COM:201:0560:FIN:EN:PDF> (accessed 10 October 2020).

296 P7_TA(2010)0352, EU action on oil exploration and extraction in Europe, European Parliament resolution of 7 October 2010 on EU action on oil exploration and extraction in Europe <http://www.europarl.europa.eu/sides/getDoc.do?pubRef=-//EP//NONSGML+TA+P7-TA-2010-0352+0+DOC+PDF+V0//EN> (accessed 10 October 2020).

Union waters and to limit the consequences of such an accident, by supplementing the acquis with a set of specific rules. The aim is therefore to promote, for the first time in European Union Law, a comprehensive legislative framework with the objective of guaranteeing the highest safety standards in the world. In this perspective, the Commission made the initial choice of a regulation, directly binding for the Member States, by publishing on 27 October 2011, a Proposal for a Regulation of the European Parliament and of the Council on safety of offshore oil and gas prospection, exploration and production activities.[297] This option reflects the Commission's ambition and willingness to establish a harmonized European regime that is automatically incorporated into the national law of all EU Member States.

However, this initial strategy naturally had to face opposition from States already involved in offshore operations, hostile to the adoption of a regulation. At their initiative and with the active support of the oil and gas lobbies,[298] the European Parliament's Industry Research and Energy ("ITRE") Committee, supported by the Environment, Public Health and Food Safety ("ENVI") Committee, introduced an amendment to transform the regulation into a directive.[299] While noting that "a few delegations still favour a Regulation,"[300] the Council supported this a mini-

---

297 Proposal for a Regulation of the European Parliament and of the Council on safety of offshore oil and gas prospection, exploration and production activities, Brussels, 27.10.2011, COM(2011) 688 final, 2011/0309 (COD) <http://eur-lex.europa.eu/legal-content/EN/TXT/HTML/?uri=CELEX:52011PC0688&from=EN> (accessed 10 October 2020).

298 On lobbies in the European Union, Pierre-Yves Monjal, 'Transparency and Lobbying or How Transparency According the European Legal Frame Seems To Vanish In the Haze' [2016] *American International Journal of Social Science* 104.

299 Final Report *Safety of offshore exploration and exploitation activities in the Mediterranean: creating synergies between the forthcoming EU Regulation and the Protocol to the Barcelona Convention* 14 <http://ec.europa.eu/environment/marine/international-cooperation/regional-sea-conventions/barcelona-convention/pdf/Final%20Report%20Offshore%20Safety%20Barcelona%20Protocol%20.pdf> (accessed 10 October 2020).

300 Council of the European Union, Proposal for a Regulation of the European Parliament and of the Council on safety of offshore oil and gas prospection, exploration and production activities - Presidency report on state of play, Brussels, 22 November 2012, Interinstitutional File: 2011/0309 (COD), 16604/12, 2, § 4 a) <http://register.consilium.europa.eu/doc/srv?l=EN&f=ST%2016604%202012%20INIT> (accessed 10 October 2020).

ma option under the influence of the States most concerned by offshore activities, first and foremost Denmark, the Netherlands and the United Kingdom; and the Commission's initial strategy was definitively disavowed at the Transport, Telecommunications and Energy Council on 3 December 2012. In this context, the Parliament and the Council reached an agreement on 21 February 2013 and recommended the adoption of a directive, a disappointing solution because it is much less ambitious since it only establishes the objectives, i.e., an obligation of result for the States, but leaves them completely free as regards the means to implement them, during the process of transposition of the directive into national law.

However, the Material Framework: Safety of Offshore Oil and Gas Operations can also be considered a disappointment.

3.3.2.1.2 Material framework: Safety of Offshore Oil and Gas Operations

Indeed, the material scope of the Directive is actually quite limited,[301] even though it is an act adopted under Title XX Environment, and in particular Articles 191 and 192 § 1.[302] Environmental concerns are certainly not absent, even if they seem in the end to have been relegated to the background and are unfortunately among those which remain the vaguest and least prescriptive.[303]

---

301 The synergy with the Offshore Protocol is real but only partial, and the comparison is not in favor of the Directive, which in theory has a broader scope and belongs to another normative generation, but in practice proves to be much less ambitious in material terms. On this aspect, see Final Report *Safety of offshore exploration and exploitation activities in the Mediterranean: creating synergies between the forthcoming EU Regulation and the Protocol to the Barcelona Convention* <http://ec.europa.eu/environment/marine/international-cooperation/regional-sea-conventions/barcelona-convention/pdf/Final%20Report%20Offshore%20Safety%20Barcelona%20Protocol%20.pdf> (accessed 10 October 2020); and for a doctrinal analysis, Nathalie Ros, 'La réglementation euro-méditerranéenne des activités offshore' [2015] *Diritto del Commercio Internazionale* 93, esp. 127-130.

302 Both are mentioned in the Preamble.

303 Article 4 *Safety and environmental considerations relating to licences*: '6. When assessing the technical and financial capabilities of an applicant for a licence, special attention shall be paid to any environmentally sensitive marine and coastal environments, in particular ecosystems which play an important role in mitigation and adaptation to climate change, such as salt marshes and sea grass beds, and marine protected areas, such as special areas of conservation pursuant to the Council Directive 92/43/EEC of 21 May 1992 on the conservation of natu-

Developed from an economic perspective and essentially centered on a logic of industrial safety, the objective of the Directive is articulated around three main axes: to reduce as far as possible the occurrence of major accidents relating to offshore oil and gas operations and; to limit their consequences, thus increasing the protection of the marine environment and coastal economies against pollution, establishing minimum conditions for safe offshore exploration and exploitation of oil and gas and limiting possible disruptions to Union indigenous energy production, and; to improve the response mechanisms in case of an accident.[304] However, under Article 1 § 1, "this Directive establishes minimum requirements for preventing major accidents in offshore oil and gas operations and limiting the consequences of such accidents"; and the improvement of workers' environmental and health conditions does not therefore appear explicitly under the Subject and scope[305] of the Directive, although it contributes to it indirectly insofar as those minimum requirements are such as to promote it. However, only the accidental dimension of the offshore activity is taken into account, and operational pollution and its effects therefore remain outside the scope of the Directive.

From a spatial point of view, the Directive applies "offshore," i.e., according to Article 2 § 2, "in the territorial sea, the Exclusive Economic Zone or the continental shelf of a Member State within the meaning of the United Nations Convention on the Law of the Sea." In other words, its geographical scope is limited to "marine waters covered by the sovereignty or sovereign rights and jurisdiction of Member States [in] the four marine regions identified in Article 4(1) of Directive 2008/56/EC, namely the Baltic Sea, the North-east Atlantic Ocean, the Mediterranean Sea and the Black Sea" (Recital 50). Although logical, by virtue of the principle of the relative effect, this solution is of course not entirely satisfactory for combating phe-

---

ral habitats and of wild fauna and flora, special protection areas pursuant to the Directive 2009/147/EC of the European Parliament and of the Council of 30 November 2009 on the conservation of wild birds, and marine protected areas as agreed by the Union or Member States concerned within the framework of any international or regional agreements to which they are a party.'

304 Recital 2 of the Preamble.
305 Article 1 *Subject and scope*: '2. This Directive shall be without prejudice to Union law concerning safety and health of workers at work, in particular Directives 89/391/EEC and 92/91/EEC. 3. This Directive shall be without prejudice to Directives 94/22/EC, 2001/42/EC, 2003/4/EC, 2003/35/EC, 2010/75/EU and 2011/92/EU.'

nomena with a transboundary dimension inherent in their nature, such as pollution and the other induced effects of these "major accidents" (Article 2 § 1). This is a fortiori so insofar as personal jurisdiction does not provide satisfactory remedies, which is precisely also the case here.

From a material point of view, "'offshore oil and gas operations' means all activities associated with an installation or connected infrastructure, including design, planning, construction, operation and decommissioning thereof, relating to exploration and production of oil or gas, but excluding conveyance of oil and gas from one coast to another" (Article 2 § 3). Where the Mediterranean Protocol developed a global approach to offshore exploration and exploitation including the whole process, and all mineral resources, the Directive deals only with the safety of offshore operations and is limited only to oil and gas exploration and exploitation activities, hence the express exclusion of transport operations.

Because of its subject, scope, lack of coherence with the acquis, and also its low normative ambition, the Directive[306] is a disappointing contribution, in the form of Challenges for EU Blue Growth.

3.3.2.2 Challenges for EU Blue Growth

The 2013 Directive is presented as a contribution to the Union's Integrated Maritime Policy, in which Blue Growth is deeply rooted and involved. With regard to its environmental pillar, the 2008 Marine Strategy Framework Directive, and according to Recital 7 of the Preamble, "Directive 2008/56/EC aims to address, as one of its central purposes, the cumulative impacts from all activities on the marine environment, and is the environmental pillar of the Integrated Maritime Policy. That policy is relevant to offshore oil and gas operations as it requires the linking of particular concerns from each economic sector with the general aim of ensuring a comprehensive understanding of the oceans, seas and coastal areas, with the objective of developing a coherent approach to the seas taking into account all economic, environmental and social aspects through the use of maritime spatial planning and marine knowledge."

In practice, however, the Directive is a Limited Normative Contribution with a Relative Legal Applicability.

---

306 José Juste Ruiz, 'La directive européenne sur la sécurité des opérations pétrolières et gazières en mer' [2014] *Revue Juridique de l'Environnement* 23.

3.3.2.2.1 A Limited Normative Contribution

Indeed, from a normative point of view, the final result appears far below initial expectations. In parallel with the adoption of a directive instead of a regulation, and as usual, the oil and gas lobbies and the majority of States with a very active offshore industry, including Denmark, the Netherlands and the United Kingdom, have succeeded in reducing the normative scope and legal quality of the Directive. Many norms are soft law, suffer from a lack of precision or clarity due to an intentionally complex drafting, and provide partial solutions to global problems. No independent supervision for authorization, evaluation and control has been established or provided for, since none of these tasks was finally devolved to the European Maritime Safety Agency.[307] The margin of appreciation left to the States does not promote the effective safety of offshore oil and gas operations, but allows for a self-regulation very favorable to the development of an industrial sector considered to be economically promising.

This is a challenge for the legal order of the European Union, as well as for the Blue Growth perspective associated with the Integrated Maritime Policy; and this is all the more so because the Directive also has a Relative Legal Applicability.

3.3.2.2.2 A Relative Legal Applicability

Although its legal basis is the protection of the marine environment (Part XX Environment), its purpose is limited to the establishment of mini-

---

[307] As useful as it may be, the Agency's mission falls short of what was initially envisaged; Article 10 *Tasks of the European Maritime Safety Agency*: '1. The European Maritime Safety Agency (EMSA, hereinafter 'Agency') shall provide the Member States and Commission with technical and scientific assistance in accordance with its mandate under Regulation (EC) No 1406/2002. 2. Within the framework of its mandate, the Agency shall: (a) assist the Commission and the affected Member State, on its request, in detecting and monitoring the extent of an oil or gas spill; (b) assist Member States, at their request, with the preparation and execution of external emergency response plans, especially when there are transboundary impacts within and beyond offshore waters of Member States; (c) on the basis of the Member States' external and internal emergency response plans, develop with Member States and operators a catalogue of emergency equipment and services available. 3. The Agency may, if requested: (a) assist the Commission in assessing the external emergency response plans of Member States to check whether the plans are in conformity with this Directive; (b) review exercises that focus on testing transboundary and Union emergency mechanisms.'

mum requirements applicable in the event of major accidents; this is of course without prejudice to Article 193 of the Treaty on the Functioning of the European Union according to which "The protective measures adopted pursuant to Article 192 shall not prevent any Member State from maintaining or introducing more stringent protective measures. Such measures must be compatible with the Treaties. They shall be notified to the Commission."

However, from the point of view of environmental protection, which appears actually subsidiary since the Directive, unlike the Mediterranean Protocol, is not oriented towards this dimension nor towards pollution response, the definition of "major accident" seems to be extremely reductive (Article 2 § 1 (d)).[308] The drafters of the text have indeed chosen to necessarily include the human element, apprehended by reference to "fatalities or serious personal injury." The definition therefore deliberately has a limiting effect, of such a nature as to undermine its applicability, since, assessed in the light of that criterion, accidents such as those of the Erika or Prestige could not have been regarded as major accidents.

The Directive must of course be transposed, but it is in fact a transposition with variable geometry, *ratione temporis*[309] and *ratione perso-*

---

308 Article 2 *Definitions*: 'For the purpose of this Directive: (1) 'major accident' means, in relation to an installation or connected infrastructure: (a) an incident involving an explosion, fire, loss of well control, or release of oil, gas or dangerous substances involving, or with a significant potential to cause, fatalities or serious personal injury; (b) an incident leading to serious damage to the installation or connected infrastructure involving, or with a significant potential to cause, fatalities or serious personal injury; (c) any other incident leading to fatalities or serious injury to five or more persons who are on the offshore installation where the source of danger occurs or who are engaged in an offshore oil and gas operation in connection with the installation or connected infrastructure; or (d) any major environmental incident resulting from incidents referred to in points (a), (b) and (c). For the purposes of determining whether an incident constitutes a major accident under points (a), (b) or (d), an installation that is normally unattended shall be treated as if it were attended.' Major environmental incident' is cumulatively defined in Article 2 § 37, as 'an incident which results, or is likely to result, in significant adverse effects on the environment in accordance with Directive 2004/35/EC.'

309 As regards the transposition, 'Member States shall bring into force the laws, regulations, and administrative provisions necessary to comply with this Directive by 19 July 2015' (Article 41 § 1). In relation to owners and operators, a distinction must be made depending on whether the installations are planned or existing; a period of three years, by 19 July 2016, is provided for the operational

nae.[310] Indeed, only coastal Member States of the Union, under whose jurisdiction offshore operations are carried out, are required to transpose the entire Directive. Other coastal Member States are only required to transpose the two articles on transboundary emergency situations (Article 32)[311] and penalties (Article 34),[312] as well as Article 20, Offshore oil and gas operations conducted outside the Union, referring to the provision of a report in the event of a major accident occurring during operations outside the EU (Article 41 § 3). This article is, moreover, the only one which landlocked States are also obliged to transpose (Article 41 § 4). But States where no offshore company is registered are not affected by this provision, whether coastal or landlocked, and they shall be obliged to bring into force those measures which are necessary to ensure compliance with Article 20 only as from 12 months following any later registration of such a company (Article 41 § 5). This combination of provisions has been strongly criticized by the Commission, which regrets that some Member States may thus be partly exempted from the obligation to incorporate the Directive into their domestic law. In the Declaration which it attached to the Directive, the Commission underlined the risks inherent in this precedent, both for the integrity of EU Law and in terms of possible circumvention;[313] this illustrates the complexity and

---

implementation of the Directive on future installations (Article 42 § 1), but in the case where they already exist, a period of five years applies for compliance, with a deadline by 19 July 2018 (Article 42 § 2).

310 Article 41 *Transposition*.

311 Article 32 *Transboundary emergency preparedness and response of Member States without offshore oil and gas operations under their jurisdiction*.

312 Article 34 *Penalties*.

313 *Statement by the Commission*: '1. The Commission regrets that under paragraphs 3 and 5 of Article 41 some Member States are partially exempted from the obligation to transpose the Directive and considers that such derogations shall not be regarded as a precedent in order not to affect the integrity of EU law. 2. The Commission notes that Member States may use the option not to transpose and apply Article 20 of the Directive because of the current absence of any company registered in their jurisdiction which has offshore activities outside the territory of the Union. In order to ensure effective enforcement of this Directive, the Commission underlines that it is incumbent on these Member States to ensure that companies already registered with them do not circumvent the aims of the Directive by extending their business objects to include offshore activities without notification of this extension to the competent national authorities so that they can take the necessary steps to ensure full application of Article 20. The Commission will take all necessary measures against any circumvention which

difficulties inherent in the challenges associated with the effective regulation of offshore activities.

However, the relatively limited applicability of the Directive is above all patent *ratione loci*, i.e., from a spatial point of view; it is ontologically an act of European Union Law and transposition is only really necessary for offshore operations carried out in maritime areas under the jurisdiction of the Member States, since under Article 20 offshore oil and gas operations conducted outside the Union, directly or indirectly by companies registered in the territory of Member States, can only give rise to a possible report in the event of a major accident.[314]

In other words, and to speak the language of International Law of the Sea, the Directive only applies in marine waters under the sovereignty or jurisdiction of the Member States of the European Union, i.e., in the territorial sea, but in principle not in internal waters, and in the exclusive economic zone as well as on the continental shelf; it does not apply outside EU marine waters, on the high seas or in the waters of non-EU Member States, including when operators and owners are registered in the territory of a Member State.[315] This approach falls far short of what exists in other sectors of the maritime economy, for example fisheries.[316] It also appears

---

may be brought to its attention.'

314 Article 20 *Offshore oil and gas operations conducted outside the Union*: '1. Member States shall require companies registered in their territory and conducting, themselves or through subsidiaries, offshore oil and gas operations outside the Union as licence holders or operators to report to them, on request, the circumstances of any major accident in which they have been involved. 2. In the request for a report pursuant to paragraph 1 of this Article, the relevant Member State shall specify the details of the information required. Such reports shall be exchanged in accordance with Article 27(1). Member States which have neither a competent authority nor a contact point shall submit the reports received to the Commission.'

315 These questions are dealt with in a very soft way and in votive mode in Recitals 36 to 41 of the Preamble.

316 Article 1 of Regulation (EU) No 1380/2013 of the European Parliament and of the Council of 11 December 2013 on the Common Fisheries Policy, amending Council Regulations (EC) No 1954/2003 and (EC) No 1224/2009 and repealing Council Regulations (EC) No 2371/2002 and (EC) No 639/2004 and Council Decision 2004/585/EC defines its *Scope* as follows: '1. The Common Fisheries Policy (CFP) shall cover: (a) the conservation of marine biological resources and the management of fisheries and fleets exploiting such resources; (b) in relation to measures on markets and financial measures in support of

paradoxical with regard to the objective of the Directive, because a major accident in an EU marine region, but outside EU waters, can of course have serious consequences in the waters of Member States ... This type of hypothesis could, for example, occur in the Mediterranean Sea, but also in other marine regions and regional systems in which the European Union intervenes and participates as a subject of Public International Law, either directly as a member (Baltic, Atlantic), or indirectly via its Member States (Black Sea, Arctic). Marine pollution and the damage it is likely to cause to the environment, both at sea and on land, know no frontiers, neither those of States nor those of the Union.

3.4 Conclusion

Like Public International Law, European Union Law therefore appears to have failed to impose a real, binding and specific legal framework to regulate offshore energy activities, i.e., offshore oil and gas exploration and exploitation operations, while other forms of mining of the resources of the continental shelf already at work in certain regions of the world have also deliberately not been included in the Directive.[317] The development of offshore industrial activities has in fact been clearly privileged, instead

---

the implementation of the CFP: fresh water biological resources, aquaculture, and the processing and marketing of fisheries and aquaculture products. 2. The CFP shall cover the activities referred to in paragraph 1 where they are carried out: (a) on the territory of Member States to which the Treaty applies; (b) in Union waters, including by fishing vessels flying the flag of, and registered in, third countries; (c) by Union fishing vessels outside Union waters; or (d) by nationals of Member States, without prejudice to the primary responsibility of the flag State'; Article 28 § 2 (d) provides 'in particular, the Union shall: [...] (d) ensure that Union fishing activities outside Union waters are based on the same principles and standards as those applicable under Union law in the area of the CFP, while promoting a level-playing field for Union operators vis-à-vis third-country operators.' Council Regulation (EC) No 1005/2008 of 29 September 2008 establishing a Community system to prevent, deter and eliminate illegal, unreported and unregulated fishing is even more explicit; paragraph 3 of Article 1 *Subject matter and scope* provides 'the system laid down in paragraph 1 shall apply to all IUU fishing and associated activities carried out within the territory of Member States to which the Treaty applies, within Community waters, within maritime waters under the jurisdiction or sovereignty of third countries and on the high seas.'

317   Unlike the Mediterranean Protocol, the Directive is not applicable in the case of the exploitation of rare earths, certain deposits of which exist in Europe, particularly in the Mediterranean Sea.

of a genuine sustainable development approach which really takes into account the interests of workers[318] and environmental protection, in accordance with the three pillars principle, while nothing is done to regulate convenience registrations of platforms.[319] However, the Integrated Maritime Policy refers to a global approach to all European Union policies relating to the sea, which is in line with the postulate set out in the Preamble to the United Nations Convention on the Law of the Sea, according to which "the problems of ocean space are closely interrelated and need to be considered as a whole," with objectives in principle very close to those of sustainable development and its threefold dimension.[320]

Blue Economy and Blue Growth, which are an integral part of the Integrated Maritime Policy and UN programming strategies, are therefore more than ever showing their true face,[321] that of a misleading concept in the reference made to sustainable development goals,[322] responsible use

---

318 The human dimension of platform disasters must always be kept in mind: 11 people lost their lives in the *Deepwater* explosion in 2010, 167 in that of the *Piper Alpha* in 1988, 123 in the overturning of the *Alexander Kielland* platform in 1980, etc.

319 For example, the *Deepwater* platform, owned by the world leader in offshore, the Swiss company *Transocean*, was registered in the Marshall Islands.

320 The 2030 Agenda for Sustainable Development adopted by the UN General Assembly on 25 September 2015, included among its 17 sustainable development goals, Goal 14 (SDG 14) *Conserve and sustainably use the oceans, seas and marine resources for sustainable development*, as well as nine other objectives also presented as connected to oceans and coastal issues (*End poverty in all its forms everywhere* (SDG 1); *End hunger, achieve food security and improved nutrition and promote sustainable agriculture* (SDG 2); *Ensure healthy lives and promote well-being for all at all ages* (SDG 3); *Ensure availability and sustainable management of water and sanitation for all* (SDG 6); *Ensure access to affordable, reliable, sustainable and modern energy for all* (SDG 7); *Promote sustained, inclusive and sustainable economic growth, full and productive employment and decent work for all* (SDG 8); *Take urgent action to combat climate change and its impacts* (SDG 13); *Protect, restore and promote sustainable use of terrestrial ecosystems, sustainably manage forests, combat desertification, and halt and reverse land degradation and halt biodiversity loss* (SDG 15); *Strengthen the means of implementation and revitalize the Global Partnership for Sustainable Development* (SDG 17)).

321 For an analysis specifically dedicated to the African continent, Nathalie Ros, 'Le continent africain face aux mirages de la croissance bleue' in Jean-Baptiste Harelimana (ed.), *Liber Amicorum Stéphane Doumbé-Billé Autour du droit international économique en Afrique* (African Academy of International Law Practice 2021) forthcoming.

322 Nathalie Ros, 'Développement durable et droit de la mer' [2017] *Annuaire de*

of the resources of the sea and respect for the environment, to appear exclusively in the economic dimension of their exploitation.[323]

Bibliography

1. Marie Bourrel, 'L'Union européenne adhère au protocole sur les activités offshore en Méditerranée' [2013] Droit de l'environnement 212.

2. José Juste Ruiz, 'La directive européenne sur la sécurité des opérations pétrolières et gazières en mer' [2014] Revue Juridique de l'Environnement 23.

3. John Warren Kindt, 'The Law of the Sea: Offshore Installations and Marine Pollution' [1985] Pepperdine Law Review 381.

4. Judith van Leeuwen, Who greens the waves? Changing authority in the environmental governance of shipping and offshore oil and gas production (Wageningen Academic Publishers 2010).

5. Nengye Liu, 'The European Union's Potential Contribution to Enhanced Governance of Offshore Oil and Gas Operations in the Arctic [2015] Review of European, Comparative and International Environmental Law 223.

6. Diethard Mager, 'Climate Change, Conflicts and Cooperation in the Arctic: Easier Access to Hydrocarbons and Mineral Resources?' [2009] The International Journal of Marine and Coastal Law 347.

7. Vicente Marotta Rangel, 'Le plateau continental dans la Convention de 1982 sur le droit de la mer' [1985] Recueil des cours de l'Académie de droit international de La Haye 269.

8. Cécile Pelaudeix, 'Governance of Arctic Offshore Oil & Gas Activities: Multilevel Governance & Legal Pluralism at Stake' [2015] Arc-

---

Droit Maritime et Océanique 147; Nathalie Ros, 'Sustainable Development Approaches in the New Law of the Sea' [2017] *Spanish Yearbook of International Law* 11.

323 On correlative deviances, in terms of privatization of maritime spaces and marines resources, Nathalie Ros, 'Estrategia *Blue Growth* y retos de privatización del mar' in Jaime Cabeza Pereiro y Belén Fernández Docampo (Coord.), *Estrategia Blue Growth y Derecho del Mar* (Editorial Bomarzo 2018) 227.

tic Yearbook 214.

9. Jean-François Pulvenis, 'Le plateau continental. Définition et régime des ressources' in René-Jean Dupuy et Daniel Vignes (Dir.), Traité du nouveau droit de la mer (Economica Bruylant 1985) 275.

10. Evangelos Raftopoulos, 'Sustainable Governance of Offshore Oil and Gas Development in the Mediterranean: Revitalizing the Dormant Mediterranean Offshore Protocol' (2010) MEPIELAN E-Bulletin.

11. <http://www.mepielan-ebulletin.gr/default.aspx?pid=18&CategoryId=4&ArticleId=29&Article=Sustainable-Governance-of-Offshore-Oil-and-Gas-Development-in-the-Mediterranean:-Revitalizing-the-Dormant-Mediterranean-Offshore-Protocol> accessed October 10, 2020.

12. Nathalie Ros, 'Exploration, Exploitation and Protection of the Mediterranean Continental Shelf' in Claudia Cinelli and Eva Maria Vásquez Gómez (Ed.), Regional Strategies to Maritime Security: a Comparative Perspective (Tirant lo Blanch 2014) 101.

13. Nathalie Ros, 'La réglementation euro-méditerranéenne des activités offshore' [2015] Diritto del Commercio Internazionale 93.

14. Nathalie Ros, 'Problems of Marine Pollution resulting from Offshore Activities according to International and European Union Law', in Andrea Caligiuri (Ed.), Offshore Oil and Gas Exploration and Exploitation in the Adriatic and Ionian Seas (Editoriale Scientifica 2015) 34

15. Nathalie Ros, 'Quel régime juridique pour l'exploitation offshore en Méditerranée ?' [2015] Annuaire de Droit Maritime et Océanique 205.

16. Nathalie Ros, 'Vers une gouvernance régionale de l'offshore en mer Méditerranée ?' in Angela Del Vecchio and Fabrizio Marrella (Dir.), International Law and Maritime Governance. Current Issues and Challenges for the Regional Economic Integration Organizations / Droit international et gouvernance maritime. Enjeux actuels et défis pour les organisations régionales d'intégration économique / Diritto Internazionale e Governance Marittima. Problemi Attuali e Sfide per le Organizzazioni di Integrazione Economica Regionale, Cahiers de

l'Association internationale du Droit de la Mer 3 (Editoriale Scientifica 2016) 219.

17. Nathalie Ros, 'Environmental Challenges of Offshore Activities in International and European Union Law' in Andrea Caligiuri (Ed.), Governance of the Adriatic and Ionian Marine Space, Cahiers de l'Association internationale du Droit de la Mer 4 (Editoriale Scientifica 2016) 203.

18. Nathalie Ros, 'Retos de la Política Marítima Integrada de la Unión Europea frente a las perspectivas de explotación mar adentro' in Laura Carballo Piñeiro (Coord.), Retos presentes y futuros de la Política Marítima Integrada de la Unión Europea (Editorial Bosch 2017) 329.

19. Nathalie Ros, 'La pollution résultant de l'exploitation du sol et du sous-sol : le cas du plateau continental' in Droit des sites et sols pollués. Bilans et perspectives (L'Harmattan 2018) 39.

20. Nathalie Ros, 'L'Union européenne et l'encadrement juridique des activités offshore' [2018] Revue du droit de l'Union européenne 233.

21. Nathalie Ros, 'L'Etat côtier et son plateau continental : enjeux et perspectives dans le nouveau droit de la mer' in Liber Amicorum Haritini Dipla (Pedone 2020) 109.

22. Tullio Treves, 'La pollution résultant de l'exploration et de l'exploitation des fonds marins en droit international' [1978] Annuaire français de droit international 827.

23. Sergei Vinogradov, 'The Impact of the Deepwater Horizon: The Evolving International Legal Regime for Offshore Accidental Pollution Prevention, Preparedness, and Response' [2013] Ocean Development & International Law 335.

4

THE PLAYERS IN THE NEW ENERGY SYSTEM: WHAT ROLE FOR THE STATE IN THE ANTHROPOCENE ERA?

Aubin Nzaou-Kongo

> Energy is necessary for daily survival. Future development crucially depends on its long-term availability in increasing quantities from sources that are dependable, safe, and environmentally sound. At present, no single source or mix of sources is at hand to meet this future need.

4.1 Introduction

Energy is an essential good that provides the most primary services for human life.[324] The common future of mankind rests on a steady availability of energy from various sources.[325] However, the current energy mix is overwhelmingly composed of non-renewable energy,[326] which requires a paradigm shift in order to meet humanity's future needs[327] and to achieve energy efficiency.[328] Hence, within two centuries—suffice it to say—all countries became dependent on fossil fuels.[329] By virtue of a long-standing colonial legacy, African states have ironically entered the "fossil fuel civilization."[330] As a result, they claimed their right to development, joined the peloton of the industrial civilization—[331] at their own pace and steady effort—and chased the promise of

---

324 World Commission on Environment and Development, *Our Common Future* (Oxford University Press 1987) 168.

325 V. Smil, *Energy: a Beginner's Guide* (Oneworld Publications 2006) 127.

326 V. Smil, *Energy and Civilization: A History* (The MIT Press, 2017) 295 (affirming the current share '2015' of fossil fuels (86%) in global primary energy, which is only 4% less than a generation ago '1990').

327 UN Environment (ed.), *Global Environment Outlook – GEO-6: Healthy Planet, Healthy People* (Cambridge University Press 2019) 9.

328 Renewable Energy Policy Network for the 21st Century [REN21], *Renewables 2018: Global Status Report*, (REN21 2018) 165.

329 V. Smil (n 328) 295-96.

330 Ibid 228.

331 V. Smil, *The Earth's Biosphere: Evolution, Dynamics, and Change* (The MIT Press

development. In this sense, they disregarded their initial natural harmony and waded in a constraining, destructive and self-defeating venture.[332] Above all, they have still had to bear disproportionately the double burden of the social and economic impacts of the Industrial Revolution over time.[333] To their detriment, they have also been subjected to a persistent environmental and climate ordeal.

These challenges are so enormous for African states that they require a timely action. Earlier this decade, over half of the African people was without access to energy services. That is, nearly 590 million Africans (57% of the population) could not afford or lived in off-grid areas.[334] A sharp increase is observed when access to clean cooking fuels and cooking technologies is added to this figure, augmenting in the vicinity of 700 million people (68% of the population).[335] It dawns upon everyone—with an unambivalent conclusiveness—that Africa is a widely diverse continent. Such diversity translates into varying levels of development between states, including the specific place of the energy sector in the national economic landscape. Thereby, fossil fuels and renewable energies are not equally incorporated into energy policies, which arguably leads to unequal access to energy services and soars the energy poverty. Consequently, this situation necessarily impacts—over the course of time—the overall productive activity and the access and consumption of communities and households. This is in fact the challenge that lies ahead for the new energy system when it comes to African states.[336]

Without a sudden fright, the state has a primary and fundamental role

---

2002) 231.

332 L. Rajamani and J. Peel (eds.), *The Oxford Handbook of International Environmental Law* (Oxford University Press 2ed. 2021) 1.

333 A legitimate question that arises today is: was the industrial revolution necesary? Economists have been focusing on new questions, exploring new aspects based on new data to challenge conventional wisdom. G. SNOOKS, WAS THE INDUSTRIAL REVOLUTION NECESSARY? (Routledge 3ed. 2002).

334 Irena, *Africa's Renewable Future: The Pathway to Sustainable Growth* (IRENA 2013) 5.

335 Ibid.

336 G. Nzobadila, *Energy Poverty in Africa*, (African Energy Commission 2017), <https://au-afrec.org/Docs/FR/PDF/2017/paper_on_africa_energy_poverty_en.pdf> (accessed 05 July 2020).

to play here.³³⁷ It entertains no doubt that major decisions must be taken to transform the energy mix, by increasing the share of renewable energies, alongside other socio-economic issues that need to be addressed.³³⁸ Domestic consumption is proving to be a genuine intersectional issue,³³⁹ including access to energy services, economic and energy poverty, energy efficiency and conservation, and other social challenges.³⁴⁰ Although renewable energy can make a considerable contribution to the energy mix, the state must actually reduce the share of energy resources usually exported to meet domestic needs. The changing environment of African states appears relatively challenging.³⁴¹

Through this framework, I examine the role of African states in the context of the global change represented by the energy transition. Focusing on French-, Portuguese- and English-speaking states allow us to cross-reference three different legal traditions, but also to take a broader look at the conceptual meaning and materiality of the transition emerging from these different states. Most of these countries have historically experienced failed social experiments by various actors, including governments, non-state actors, and international organizations, on economic, social, and environmental issues. Although the terminology tends to evolve nowadays, not so long ago they were literally considered "undeveloped countries."³⁴² A conceptual approach that only finds real meaning in economic studies.³⁴³ One could argue ab initio that the concept of development per se is problematic.³⁴⁴ It is certainly a by-product of biological studies that has been arbitrarily applied to the trajectory of social and economic progress of states in recent years. Most of them are

---

337 T.-L. Field, 'A Just Energy Transition and Functional Federalism: The Case of South Africa' (2021) 10 *Transnational Environmental Law* 237.

338 Ibid 238.

339 T. Etty et al., 'Energy Transition in a Transnational World' (2021) 10 *Transnational Environmental Law* 199.

340 Ibid.

341 J. Saurer & J. Monast, 'The Law of Energy Transition in Federal Systems' (2021) 10 *Transnational Environmental Law* 205.

342 E. Wayne Nafziger, *Economic Development* (Cambridge University Press 4ed. 2006) 21.

343 Ibid 02.

344 Ibid 21.

in three different regions (Africa, Asia, and Latin America), facing different social and economic problems. Historically, they bear the burden of a disproportionate distribution of power and wealth,[345] resulting from industrial and economic domination.[346] However, one should be legitimate to ask: how come those underdeveloped countries could aspire to or achieve sustainable development? Could a country sustain its development while it is not developed? Can underdeveloped countries pretend to achieve sustainable development? Can sustainable development be a crossroads where developed and non-developed countries meet without any discrimination? Is sustainable development becoming the only mode of development? Does it definitively replace economic development, in the sense that achieving sustainable development means no longer going through the economic trajectory?[347]

To date, it is hard to find a solid answer to those mere questions. Yet there are still about 50 million indigenous people in Africa[348] who live in great harmony with nature, passing on ancestral knowledge from generation to generation, and—for the most part—exploiting resources and lands very rationally. The Congo Basin and the Niger Delta epitomize the large areas that are currently home to many groups. When it comes to preserving lands and natural resources, their practices and lifestyle have proven to be effective. Anyone who has had the privilege of accessing their living environment could testify to this very fact. Is their way of life sustainable enough to be preserved for our common good? Should we continue to jeopardize their living environment in the name of economic growth driven by mining, expropriation, and deforestation?[349] As one can see, they bear a double burden. First, the burden of a disharmonious way of life that has led to a global catastrophe on nature and the ecosystem. As a

---

345 A. Shorrocks and R. van der Hoeven (eds.), *Growth, Inequality, and Poverty Prospects for Pro-Poor Economic Development* (Oxford University Press, 2004) 144 (analyzing the conditions for redistribution of growth between different areas).

346 Ibid 02.

347 K. Persson, *An Economic History of Europe: Knowledge, Institutions and Growth, 600 to the present* (Cambridge University Press 4ed. 2010) 21.

348 See the International Work Group on Indigenous Affairs, <http://www.iwgia.org/regions/africa> (accessed 05 July 2020).

349 A. Jegede, 'Climate Change: Safeguarding Indigenous Peoples through 'Land Sensitive' Adaptation Policy in Africa' in Walter Leal Filho (ed.), *Handbook of Climate Change Adaptation* (Springer 2015) 799.

result, the energy transition will impose an appalling social and environmental plight on them. Secondly, the burden of the promise of economic development[350] that always benefits a small group, while the indigenous people—and part of the people—endure the hardship. Not to mention the imperiousness of those who plausibly claim to bring them sustainable development.

This article explores the significant role that the state is still expected to play in initiating and implementing the energy transition. In this regard, it is laid out in three parts. Part I focuses on the premise of the role that derives from constitutional law. This role is considered classic, because it is based on different functions of the state, and the legitimate constrain that distinguishes it from other social actors, including non-state actors. Tremendous materials are offered by the analysis either from the perspective of sociology or law studies when it comes to the specific situation of French-speaking African states. The scope of analysis is broadened with the energy law approach. With a focus on African English-speaking countries, the chapter examines both the way the state is enforcing statutes aiming to design its own transition scheme and exercising its discretionary power through its energy policy.

Beyond the functions of the state—deriving from its sovereign power—these elements set out the direction in quest of a specific role the state can play in the energy transition as a process in Part II. As such, the energy transition, if it is to lead to coherent social change, requires strong and dynamic leadership, including clear, nuanced, and forward-looking direction on the broad sections of the overall process, and the environmental justice issues that necessarily cluster around them. For this reason, the role of the state is construed as both a steering role, and an integrative role for environmental, economic, and social issues. Part III provides a rationale for the necessary and strong support of international cooperation—to the state—in order to achieve the paradigm shift smoothly. In Part IV, I emphasize the African Union's transition initiatives in the run-up to COP 25, which I hold out as an inducement for states' efforts. In fact, this chapter seeks to address these issues. Taken together, they could help build a coherent pattern of the role that African states play in the

---

350 F. Sarr, *Afrotopia* (Drew S. Burk & Sarah Jones-Boardman (trans.), University of Minnesota Press 2019) 5.

energy transition.

4.2 The Role of the State: An Old Question Asked Anew

4.2.1 A Constitutional Law Rationale

Even in its best state, the state is a "necessary evil."[351] This assumption is most closely associated with the thought of Thomas Paine.[352] It reminds us that, at the present stage of the evolution of human societies, the state is the most accomplished form of social entity, but also that it is here to stay. For the moment, it is certainly not going anywhere. Even today, it is the constitutions that determine the design and ends of the state.[353]

From the sociological vantage point, the state is naturally distinct from other organized social groups. The rulers claim and hold—on behalf of the state—the monopoly of legitimate violence[354] or coercive power.[355] Here, the regalian missions of the state operate, which suppose the power of sanctioning certain demeanors, certain social behaviors.[356] This primary statement is not merely a statement of fact, but a fundamental tenet.[357] It highlights the determining role the state has played historically,[358] and necessarily continues to play today in the organization of social relations.[359] As a result, any process of social transformation—[360]which has an impact on the very structure or functioning of the state or its set of values—can only be conceived from the moment when the state really

---

351 See T. Paine, *Common Sense* (Penguin Books-Great Ideas ed. 1776) 5.
352 Ibid.
353 Ibid 6.
354 See M. Weber, *Le Savant et le Politique* (Union Générale d'Éditions ed. 1963) 22.
355 See F. Fukuyama, *The End of History and the Last Man* (The Free Press ed. 1992) 16.
356 See M. Weber, (n 356) 86.
357 See L. Duguit, *Traite de Droit Constitutionnel* (E. de Boccard 3d ed. 1927) 534.
358 See G. Vedel, *Manuel Elementaire de Droit Constitutionnel* (Librairie du Recueil Sirey 1st ed. 1949) 99.
359 See M. Weber, *L'Ethique Protestante et l'Esprit du Capitalisme* (1st ed. 1905) 140; J. Chevallier, « L'Etat-Nation (1980) *Rev. de Droit Public* 1272.
360 See E. Durkheim, *Montesquieu et Rousseau: Precurseurs de la Sociologie* (1st ed. 1918) 115.

assumes its dominant position in the social order.³⁶¹ Although, it now has to do so alongside other social actors, especially civil society in general.³⁶² Such a position, more than any other, confers on the state the power to enact the rule of law,³⁶³ through the legislative process,³⁶⁴ but also to apply it through its administrative³⁶⁵ and jurisdictional powers.³⁶⁶

From a purely legal standpoint, functions of the state refer to different modes of exercising public power.³⁶⁷ These functions should not be confused with the missions of the state, the most relevant aspects of which can easily be summarized in a few points.³⁶⁸ It is the duty of the state to ensure the security of individuals and the nation against external forces,³⁶⁹ to uphold the rule of law and order within territorial borders, to guarantee and achieve the moral and material progress of individuals and the nation, including social, cultural and economic progress.³⁷⁰ In order to fulfill these missions, the state uses certain procedures or resorts to certain acts referred to as state functions. For the purposes of this reasoning, we will briefly consider three functions, whose foundations are constitutional in French- and Portuguese-speaking African states, namely the legislative, administrative, and jurisdictional function.³⁷¹

First and foremost, the legislative function is predominant in modern states. In this respect, nothing is more ordinary than the basic mention or

---

361   See R. Carre de Malberg, *Contribution À La Théorie Générale de l'État: Spécialement d'Après Les Données Fournies Par Le Droit Constitutionnel Français* (Librairie du Recueil Sirey 1st ed. 1920) 11; D. Armstrong et al., *Civil Society and International Governance: The Role of Non-State Actors in the EU, Africa, Asia and Middle East* (Routledge ed. 2015) 4; M. Carnoy, *The State and Political Theory* (Princeton University Press 1984) 7.

362   A. Hurrell, *On Global Order: Power, Values, and the Constitution on International Society* (Oxford University Press 2008) 6; D. Lewis and N. Kanji, *Non-Governmental Organizations and Development* (Routledge 2009) 5.

363   B. Boutros-Ghali, Le principe d'égalité des Etats et les organisations internationales, (1960) 100 RCADI 001, 22.

364   R. Carre de Malberg, 285.

365   Ibid 463.

366   Ibid 691.

367   Ibid 259.

368   Ibid 263.

369   T. Paine, (n 353) 5.

370   R. Carre de Malberg, supra note 5, at 260.

371   See L. Duguit, *Traite de Droit Constitutionnel* (E. de Boccard 3d ed. 1928) 151.

constitutional formula according to which:

> Parliament shall draft and vote the laws in a sovereign manner (...). Or Parliament shall have power to make laws (...)."[372]

Within the domestic legal system, this formula refers to legislative power—namely the capacity to vote the law—which was initially admitted as the "raison d'être" of the parliamentary institution.[373] For this reason, it is undoubtedly the most significant of the parliamentary functions since it is about making the law on behalf of the people. With due regard to this function, it is asserted that the National Assembly—in most cases—not necessarily the Parliament as a whole—shall express the sovereign will of the people.[374] By means of the legislative process, the law or statute is regarded as though it expresses the general will.[375] Consequently, the deep meaning of this postulate is that laws are derived from the will of people, as it is carried by its representatives in the parliament. That is to say—by its nature and importance—the law is originally an act of sovereignty. It carries out the presumption of embodying the will expressed by the people through their representatives.

The Constitution of Angola sustains it as follows:

> The National Assembly shall be a single house representing all Angolans, which shall express the sovereign will of the people and exercise the legislative power of the state.[376]

Once passed by Parliament, statutes become central legal instruments in the domestic legal order.[377] It should be noted—in this regard—that in most of French-speaking African states, statutes provide authorizations, impose prohibitions, and set abstentions in the actions of individuals, groups or public authorities. Similarly, the judicial review of adminis-

---

372 Algeria Const. Article 119, §2; Angola Const. Article 160, §1 (a); Benin Const. Article 79, §2.; Botswana Const. Article 86.
373 See R. Carre de Malberg, 286.
374 Angola Const. Article 160, §1 (a).
375 See R. Carre de Malberg, 288.
376 Angola Const. Article 141, §2.
377 See E. Durkheim, 115.

trative acts,[378] also called contentious of the legality,[379] which confers on laws the status of a reference norm in relation to other norms such as regulations, reinforces their importance. It is further appropriate to argue that the field of those statutes is generally broad. By the very nature of things, when a statute is enacted in an area in which the executive branch is supposed to be involved with its regulatory power, the matter becomes purely legislative.[380] This suggests that a specific procedure will be required for such a statute to be modified or revised by the executive. In other words, the matter must be declassified in the interests of the executive branch. Accordingly, the procedure will prevent the latter from undergoing legislative amendments, or any kind of standoff, on a matter which comes initially within the purview of its regulatory power, and then will have the ability to amend such a statute. Overall, there is a variety of statutes which—in addition to ordinary ones—deal with specific matters, including sometimes legislation on strategic directions aiming to define the framework for future legislative action in certain areas,[381] or programming laws designated to set objectives in other areas.[382] The delimitation of the respective domains of statutes and regulations has made it possible to establish an important borderline between these two domains.[383] Even though statutes may set out essential rules in certain matters, they somehow are limited to determining the general principles in those matters. As a matter of fact, the latter might need to be implemented through regulations.

Second, the administrative function is still that of day-to-day management of the state.[384] It actually consists of a set of decisions contributing

---

378 M. Bailly, 'L'acte réglementaire illégal et le décret du 28 novembre 1983' (1985) Rev. de Droit Public, 1513-14

379 Y. Petit, 'Les circonstances nouvelles dans le contentieux de la légalité des actes administratifs unilatéraux' (1993) Rev. de Droit Public, 1291-92.

380 G. Moyen, 'L'évolution récente de la distinction des domaines de la loi et du règlement en droit congolais' (2001) 55 Rev. Jur. et Pol. 155.

381 Loi d'orientation de l'Education nationale of 1991, art. 2. ; Loi d'orientation sur l'Education of 1999, art. 1.

382 *Loi* relative à la *programmation militaire* 2021-2025 of 2020, Article 1.

383 See P. Moudoudou, *Droit administratif congolais* (Harmattan 2003) 22.

384 M. Ondoa, 'Le droit administratif français en Afrique francophone: contribution à l'étude de la réception des droits étrangers en droit interne' (2002) 56 Rev. Jur. Et Pol., 287.

to the daily activity of such entity, including public services, execution of statutes, etc.[385] Since legislation cannot cover each material case in advance, it is then the responsibility of the administration to deal with ordinary cases and decide.[386] That is, statutes provide the agents of state with means of action that can simplify their daily activity.[387] As a result, the area of intervention of the administration is very vast,[388] since it concerns all decisions that are directly related to the interest of the state.[389]

Regulatory power is a fundamental element of the administrative function.[390] It is vested in a certain number of administrative authorities,[391] able to issue provisions of general scope by virtue of their power.[392] In some respects, regulatory power operates similarly to the power to make laws, in that the decrees issued have provisions of a general and impersonal nature, which distinguishes them from individual measures taken by the same authorities.[393] This is a power reserved for the administration, which excludes any legislative intervention. Within this framework, the adopted regulations are considered as autonomous regulations.[394] Consequently, the regulatory power cannot be exercised over matters that fall within the legislative power.[395] In addition to these autonomous regulations, the implementing regulations, which are intended to supplement statutes, contain measures that may implement the general rules laid

---

385 See R. Carre de Malberg, 463.

386 M. Lascombe, 'Le Premier Ministre, clef de voûte des institutions ? L'article 49, alinéa 3 et les autres' (1981) Rev. de Droit Public, 105-06; Pierre le Mire, La Réforme du pouvoir réglementaire gouvernementale (1981) Rev. de Droit Public 1241.

387 See R. Carre de Malberg, 463.

388 M. Khattabi, 'Les interventions du pouvoir réglementaire dans le domaine de la loi (la pratique constitutionnelle marocaine)' (1998) 52 Rev. Jur. et Pol., 227.

389 J-M. Breton, 'L'obligation pour l'administration d'exercer son pouvoir réglementaire d'exécution des lois. A propos de quelques décisions récentes du juge administratif' (1993) Rev. de Droit Public 1749.

390 See R. Carre de Malberg, 548.

391 T. Tassou, 'La 'chose publique' dans les États africains: sens et contresens' (2000) 54 Rev. Jur. et Pol. 301.

392 M. Khattabi, 227.

393 Ibid.

394 See P. Moudoudou, 26.

395 Ibid 22.

down by the legislator.³⁹⁶

Finally, the jurisdictional function is exercised in the context of the resolution of disputes or litigation regarding facts or legal issues, brought before the various courtrooms. This function therefore consists of applying or interpreting the law in contentious cases on which the judges must rule.³⁹⁷ This function has been greatly enhanced in recent developments on the ecological transition by the climate litigation.

In fact, energy law represents an area which applies these general rules deriving from constitutional law.

4.2.2 An Energy Law Perspective

The energy transition is rather a matter of energy law. Although most issues that relate to the transition process intersect with many other disciplines, energy law and policy fundamentally constitute a necessary means through which the energy transition will translate into law, regulation or policy. With this in mind, this section attempts to focus on the statutory provisions sometimes leading to a variety of design patterns used by some English-speaking states to shape the transition process, while it also seeks to walk through the public policy landscape and reveal some of the policy objectives providing the state with a discretionary power.

4.2.2.1 Statutory Arrangements to Energy Transition

Statutes play a fundamental role in the energy transition, both for the design and implementation of the new energy system.³⁹⁸ Besides the classic legislation in the energy sector,³⁹⁹ some interesting legislations are spawning transition efforts in this respect. Most of them could be seen as by-products of state commitments to the transition towards decarbonization, but also to ambitious climate goals in some way.⁴⁰⁰

---

396 Ibid 26.
397 M. Rousset, 'La justice administrative, pièce maîtresse de l'état de droit au Maroc' 54 *Rev. Jur. Et Pol.*, 3, 3-14 (2000).
398 O. Renn, F. Ulmer and A. Deckert, *The Role of Public Participation in Energy Transitions* (Elsevier Academic Press 1st ed. 2020) 72.
399 See e.g., Act No 2014-132 of 2014 on Electricity Code in Ivory Coast assented on March 24, 2014.
400 R. Leal-Arcas, 'The Transition Towards Decarbonization: A Legal and Policy

First, it is the primary responsibility of the state to determine the content of the transition process.[401] This means that rather than laying down general proclamations, or mere promises, laws must set a substantive content to the decarbonization process that in fact can be implemented.[402]

In Ghana—for example—the legislator in adopting the Renewable Energy Act in 2011 (hereinafter also referred to as the "2011 Act") specifies—from the outset—the meaning and significance the country intends to give to its transition effort.[403] In the preliminary section, the legislature sets out the object of the 2011 Act, which concentrates upon the development, management and use of renewable energy sources to produce heat and power as follows:

1. (1) The object of this Act is to provide for the development, management and utilization of renewable energy sources for the production of heat and power in an efficient and environmentally sustainable manner.

    (2) For the purpose of subsection (1), the object shall encompass

    (a) the provision of

    (i) a framework to support the development and utilization of renewable energy sources; and

    (ii) an enabling environment to attract investment in renewable energy sources;

    (b) the promotion for the use of renewable energy;

    (c) the diversification of supplies to safeguard energy security;

    (d) improved access to electricity through the use of renewable energy sources; the building of indigenous capacity in technology for renewable energy sources; public education on renew-

---

Exploration of the European Union' Queen Mary School of Law Legal Studies Research Paper No. 222/2016 <https://ssrn.com/abstract=2739911> (accessed 05 July 2020) (providing a thorough overview of transition towards decarbonization and underlining EU member states commitments).

401 A. Nzaou-Kongo, *L'exploitation des hydrocarbures et la protection de l'environnement en République du Congo. Essai sur la complexité de leurs rapports à la lumière du droit international* (Ph.D. Dissertation collection on file with University of Lyon 3 Library System, 2018) 51.

402 M. Beham, *State Interest and the Sources of International Law. Doctrine, Morality, and Non-Treaty Law* (London: Routledge 2018) 251.

403 The 2011 Renewable Energy Act, assented on December 31st, 2011.

able energy production and utilisation; and

(e) the regulation of the production and supply of wood fuel and bio-fuel.[404]

It follows from this object that such deployment of renewable energy sources must be done in an efficient and environmentally sustainable manner.[405] Thus, the transition to these sources is strongly encouraged by the law, while ensuring that an environment conducive to such deployment is promoted nationally, including through economic investments that support the sector.[406] These renewable energies are in fact all forms of energy obtained from non-exhaustible or non-depleting sources, including wind,[407] solar,[408] hydro,[409] biomass,[410] biofuel,[411] landfill gas,[412] sewage gas,[413] geothermal energy,[414] ocean energy,[415] as well as any other source of energy whose status will be determined by the Energy Minister.[416] This transition effort aims to encourage the development and use of renewable energy within the framework of the consultation platform between the government, the private sector and civil society that will be created by the Energy Commission.[417] Also, the 2011 Act necessarily emphasizes the diversity of sources needed for energy supply, which would preserve and ensure energy security at the national level. Yet this objective inevitably involves guaranteeing access to electricity from renewable energies, but also the respective contribution of the traditional grid and the share of renewable energies.

Ghana's Minister of Energy is a key player in this area. His role is broad

---

404   Renewable Energy Act in 2011, Section 1.
405   Ibid.
406   Ibid.
407   Ibid s. 2(a).
408   Ibid s. 2(b).
409   Ibid s. 2(c).
410   Ibid s. 2(d).
411   Ibid s. 2(e).
412   Ibid s. 2(f).
413   Ibid s. 2(g).
414   Ibid s. 2(h).
415   Ibid s. 2(i).
416   Ibid s. 2(j).
417   Ibid s. 2(b).

and includes the design, implementation, and monitoring of the national energy policy, as well as the overall supervision of Ghana's energy sector agencies. In this regard, the Minister of Energy provides the necessary policy direction, on behalf of the government, to achieve the target set by the 2011 Act.

It is therefore up to the state to set out national targets for the use of renewable energy resources. In this regard, a typical example is to be found in Gambia's Renewable Energy Act of 2013 (hereinafter also referred to as the "2013 Act")[418] which illustrates such a goal. The 2013 Act could be considered realistic because of the targets it sets, which could arguably provide an overview of the new energy mix and allow for monitoring over time. The Gambian Minister of Energy is also a key player in this area, as it is his responsibility to propose medium and long-term national targets for the use of renewable energy resources in electricity generation.[419] If circumstances require, targets could be proposed on the basis of geographical or social criteria such as diversity. Accordingly, the Gambian Minister of Energy is required to report annually to the Cabinet on these targets.[420]

For the purposes of fostering the transition, the 2013 Act also states that the Ministry determines, on the one hand, the equipment eligible for tax exemption,[421] and carries out, on the other hand, an impact study of the use of biomass for energy needs.[422] Among other things, the 2013 Act lays out The Public Utilities Regulatory Authority responsibilities, including the management of the Renewable Energy Fund[423] and formulation of the Feed-In-Tariff Rules.[424] Besides, it specifies functions the Responsible Network Utility has when it comes to determine the "safety and technical capability of the grid to connect electricity generated from renewable energy resources (…)," but also whether it will be necessary to develop new grids.[425]

---

418   Renewable Energy Act of January 1st, 2013.
419   Ibid s. 4(1)
420   Ibid s. 4(2)
421   Ibid s. 3(b)
422   Ibid s. 3(c)
423   Ibid s. 2(a)
424   Ibid s. 2(b)
425   Ibid s. 3 (a)

Secondly, it is my contention that the state has the sole power to create public bodies, whose purpose is to support the policy direction set by the Government. This is the case of the Mauritius Renewable Energy Agency ("MARENA") for example. MARENA was created as a body corporate by the Mauritius Renewable Energy Agency Act in 2015 (hereinafter also referred to as the "2015 Act").[426] Its purpose is to promote the development and use of renewable energy in Mauritius.[427] To realize the scope of MARENA's mission, it must be indicated that the 2015 Act correlates the promotion of renewable energy with the achievement of sustainable development goals.[428] In this sense, in addition to increasing the share of renewable energy in the national energy mix, MARENA will promote partnership, cooperation and networking—at different levels—with various actors in the renewable energy sector.[429] In this capacity, MARENA shall design—every five years—a national renewable energy strategic plan, but also set various instruments of soft law, including guidelines and standards for renewable energy projects, and criteria for evaluation and approval of on-grid and off-grid renewable energy projects.[430]

Most of these acts are implemented through various regulations. However, the main tool used by states is now public policies, which give them broad discretionary power and leverage. These policies are either energy and climate policies or sectoral policies, including environmental or natural resources policies.

4.2.2.2 Policy Objectives for Energy Transition

This aspect is intended to amplify the scope of this section, and—for this reason—I will try to provide a very concise overview of the state's role in shaping the transition through energy and climate policy. The spectrum of energy policies is generally always broader than that of laws and regulations. It is further argued that, in light of the separation of powers, policy matters fall under the exclusive competence of the executive branch. Therefore, in order to set policies in areas of energy and climate, the latter

---

426 Mauritius Renewable Energy Agency Act No. 11 of 2015, assented on October 2nd, 2015, and proclaimed on December 26, 2015.
427 Ibid.
428 Ibid s. 4(a).
429 Ibid s. 4(f).
430 Ibid s. 5(g).

exerts a broad discretion.

A brief discussion of energy policy in Tanzania—for example—provides the general context for policy design. The new national energy policy enacted in 2015 provides a history of the energy sector in the national economy, and the general context for the national energy sector, including electricity and oil and gas.[431] The policy addresses various issues such as local content, technology availability, financing options, etc.[432] Its purpose is—inter alia—to facilitate access to modern energy services through the development and expansion of energy infrastructure,[433] create favorable conditions in order that energy resources will help to meet domestic energy demand,[434] and encourage the diversification of energy mix through energy alternatives or renewable energies.[435] It is also worth noting that the main objective of the national energy policy is:

> To provide guidance for sustainable development and utilization of energy resources to ensure optimal benefits to Tanzanians and contribute towards transformation of the national economy.[436]

Moreover, in pursuit of this objective, the policy lays out broad principles for universal access by Tanzanians to reliable, affordable, safe, efficient and environmentally sound modern energy services.[437]

Given that, in the first place, it is the scope of the energy policy to reflect the current energy mix of the state, Zimbabwe's National Energy Policy 2012, for example, deserves careful consideration.[438] It seems correct to mention—at the outset—that Zimbabwe's National Energy Policy is well designed, and its components reflect intelligently the national mix. The policy makes a direct link between energy and development, before providing an overview of the general context in which it is approved, includ-

---

431 National Energy Policy of 2015.
432 Ibid.
433 Ibid 8-9.
434 Ibid.
435 Ibid.
436 Ibid 10.
437 Ibid.
438 National Energy Policy of 2012.

ing the state's energy situation and its contribution to the national economy.[439] Having this last point in common with Tanzania's energy policy, Zimbabwe's energy policy differs when it comes to the content of the energy mix. The latter clearly determines the respective shares of electricity,[440] fossil fuels,[441] coal,[442] nuclear[443] and renewable energies.[444] As with each sub-sector, the policy considers many relevant aspects, including the social context, policy challenges, objectives, measures, strategies, etc. As a consequence, the Ministry of Energy and Power Development is responsible, on behalf of the Government, to design, monitor performance, and regulate the energy sector.[445]

It should therefore be noted that Zimbabwe's National Energy Policy put forward the gender issues in the energy sector. For this reason, the policy predicates that the role and participation of women in the energy sector still need to be adequately addressed. In fact, some problems are identified in this area, including public or private energy infrastructure provisions that are not considered gender neutral, gender disparities, female-headed households and energy poverty, women's micro-enterprises that are by nature heat intensive but face a lack of energy supply and, especially in rural areas, of support for women who use biomass energy for cooking and heating, etc.[446]

Thus, the energy policy aims to foster goals as follows:

- Ensure that the challenge of gender equality in the energy sector becomes a visible and key concern at the policy level.

- Ensure that all energy interventions create opportunities for women's empowerment and gender equality at the programme level.

- Ensure that space and opportunities are available to women at

---

439 Ibid 1-2.
440 Ibid 8.
441 Ibid 14.
442 Ibid 20.
443 Ibid 32.
444 Ibid 22-31.
445 Ibid s. 3.
446 Ibid 51-53.

the organizational level.

- Encourage greater enrolment of women in energy-related disciplines.

In order to understand the adequate application of these measures, it is not enough to be concerned exclusively with the existence of the energy policy or the laws. The role of the state is not limited to the adoption of these instruments and to consider somehow that the transition process is underway. Its role goes beyond the mere adoption of instruments, since it is on the ground of their implementation that it can be effectively verified.

4.3 Towards an Effective Transition Effort?

The state has a tremendous role to play in the governance of the energy transition.[447] This role must be decisive.[448] Essentially, this is the mindset which is emerging through the international effort when it comes to transitioning away from fossil fuels. Over time, a key feature of that effort has been to ensure access to reliable, sustainable and modern energy services at an affordable cost. Pursuant to paragraph 8 of the preamble of resolution 71/223—as adopted by the United Nations General Assembly on December 21, 2016—it is the primary responsibility of the state to ensure access to energy services. As a result, a twofold component is deriving from this, including providing impetus and direction, and—in a rather meaningful way—reconciling transition goals with basic social stakes.[449]

4.3.1 *The Role of Steering the Transition*

The state is called upon to play a general steering role. This role stems from its primary responsibility to ensure its own economic, social and—

---

447 A. Nzaou-Kongo, 51.

448 R. Lyster and A. Bradbrook, *Energy Law and the Environment* (Cambridge University Press 2006) 78.

449 J. Ebbesson, 'Introduction: dimensions of justice in environmental law' in J. Ebbesson and Ph. Okowa (eds.), *Environmental Law and Justice in Context* (Cambridge University Press 2009) 2.

even more— environmental progress.[450] Therefore, it is incumbent upon the state to design and implement an adequate energy policy, but—above all—to integrate—into its public policies—the necessity to ensure access to reliable, sustainable and modern energy services. Here, the issue of affordability for the public also comes into play, and we will focus on this thereafter. Indeed, the decisive role lies in the direction it provides, ensuring that national development policies and strategies make access to energy a priority objective. In this regard, it is important that the state—at all levels—creates instrumental conditions for achieving its new climate and energy goals, built on traditional pillars of sustainable development.

In essence, African states face a multifaceted role. A key aspect of this role is the capacity of the state to mobilize necessary financial resources to support the transformation of its energy mix. In this sense, it must create and strengthen institutional capacities in a way that better adjust to the transition objective. Without a doubt, this implies that these capacities contribute to set a favorable context to access to more environmentally friendly technologies. This orientation is in fact correlated to the targets of sustainable development objective No. 7, promoted by the General Assembly resolution. 70/1, on September 25, 2015 ("the 2030 Agenda").[451] Nourishing the ambitious purpose of carrying the new Development Agenda beyond 2015, this new pillar of sustainable development stresses the importance for states to find crucial financial resources to achieve sustainable development goals ("SDGs").

The resolution not only puts forth a new pattern for development through a set 17 goals and 169 targets—which the United Nations ("UN") is committed to achieving in full by 2030—it also assigns specific tasks to states. In this regard, many African states are trying to align their regulatory or policy frameworks with SDGs. At the national level—one example that

---

450 U. Outka, 'Environmental Justice in the Renewable Energy Transition' (2012) 19 *J. Envtl. & Sustainability L.* 64; J. Dernbach, P. Salkin and D. Brown, 'Sustainability as a Means of Improving Environmental Justice' (2012) 19 *J. Envtl. & Sustainability L.* 11.

451 According to its official title: 'Transforming our world: The 2030 Agenda for Sustainable Development.' This program carries the ambition to eliminate all forms of poverty, achieve sustainable development and promote access to energy for all. One year after this resolution, notably on December 21, 2016, the General Assembly enshrined this goal #7 through its resolution #71/233, with identical wording.

comes immediately in mind is the Republic of Congo ("RC") Strategy for the development of the electric power, water, and sanitation sectors. We will focus on this specific policy in order to take a comprehensive view of universal access to energy services. Approved by Decree No. 2010-822 on December 31, 2010 ("the 2010 Decree"), the Strategy for the Development of the Congolese Energy Sectors affirms—from its preamble—the RC's commitment to comply with the Millennium Development Goals ("MDGs"), as now superseded by SDGs above-mentioned, as well as the decisive role that the state will have to play in the energy sector.[452]

Yet—whatever substance might have this conceptual role for African states—it necessarily calls for a threefold observation.

First, comes into play the issue of the existence or lack of adequate infrastructure for the provision of energy services. At first glance, this could be a basic ordeal when considering the possibility of providing access to energy services for all. In some cases, the lack of adequate infrastructure should be a watershed, thus call out for a breakthrough reform. The question then is how to get rid of old energy infrastructure—most specifically—that is out of use. But also, how to reorganize or recycle the existing energy infrastructure, and use it for the new goal of adapting the energy mix. The challenge is indeed not to stagger backwards, but rather to cope with the necessity to identify, set aside and either renovate or change or transform. The provision of modern and sustainable energy services to all people in some francophone African states tends rather to encourage the development of more appropriate infrastructure and technologies to facilitate people's access to energy services. Such infrastructure will help increase the share of new and renewable energy in the national energy mix. Also, it will improve the level of energy efficiency and provide the desired energy performance. In addition, such infrastructure could also lead to energy savings. Adversely, from the social standpoint, these various elements raise questions about the means that these states could have at their disposal, in order to rapidly bring about the emergence of sustainable, modern and affordable energy services. Moreover, national policies must be more favorable to this transition, notably through a commitment

---

[452] Development Strategy for the Electric Power Decr. No. 2010-822 (31 December 2010).

to establish a more equitable and sustainable system,[453] to create an environment conducive to investments in the energy system and to encourage public-private partnerships in the energy sector.

A second observation—apparently not unrelated to the previous one—concerns the organization of a climate conducive to the promotion of energy efficiency and the use of new and renewable energies. Such context that would encourage the implementation of public policies to rationalize fossil fuel subsidies, which are considered a source of waste, but also research and development in this area, including tax and customs incentives.[454] Moreover, it includes the need to reconcile the orientation towards renewable energies with the rational use of energy in general, through the use of less polluting, low-carbon techniques in the exploitation of fossil resources, as well as the consideration of sustainability in traditional energies. That said, it is also agreed that it would be advisable to increase national renewable energy production capacities, the general consequence of which would undoubtedly be the creation of new jobs.

The third observation—which is also of great importance here—regards the adaptation of national legislation to the need to guarantee access to energy services for all. This means—in the first place—integrating this new dynamic of transition into the legal instruments applicable to various fields, including construction, public procurement, electricity, industry, transport, sanitation, etc. In the absence of such an option, recourse to special and transversal legislation is not to be excluded. Also, the long lingering question of a right-to-energy services must be addressed in order to strengthen the legal standing and give materiality to this ambition.

These remarks could also have the advantage of being part of the dynamics of the response to climate change. Therefore, it is also worth mentioning ensuring access for all to energy services, as formulated in the General Assembly resolution 71/233, also includes the need to involve all national actors, and in particular local actors, to better adapt to the ever-roughening realities faced by citizens. This is a very important aspect that could

---

453 J. Stephens, E. Wilson and T.Peterson, *Smart Grid (R)Evolution: Electric Power Struggles* (Cambridge University Press 2015) 405.

454 R. Mann, 'Delivering Energy Policy in the US: the Role of Taxes' in Raphael J. Heffron and Gavin F. M. Little, (eds.), *Delivering Energy Law and Policy in the EU and the US* (Edinburgh University Press 2016) 261-62.

ensure that energy services are more adapted to the needs of public and their cost is sufficiently affordable so that they do not create important social contrasts. In fact, this approach can easily be seen as putting sustainable development into perspective in the context of trade-offs that state is called upon to make.

A proper way to embrace this challenge—it is safe to say—is to spur all actions that respond to justice and fairness. I now turn to the grounds relied on by those who find it decent to advocate for an inclusive transition.

4.3.2 The Role of fostering the Energy Justice

The reference to energy transition has tremendously overshadowed the issue of energy justice.[455] The shift it requires—by its very nature—seems far more likely to rouse close attention than any problems posed by the transition per se. Yet, the energy transition—like any major change—inevitably poses many social, economic, and even environmental problems. It goes without saying that—about most of African states—the lingering question is how undeveloped countries could aspire to and achieve sustainable development? One must assume here the close tie between energy transition and sustainable development.[456] Africa contributes the least of any continent to climate change, it still has to and will carry a huge climate change burden.[457] Sub-Saharan Africa—already grappling with the challenge of electrification in different places—has become a climate change hotspot.[458] As a result, it will inescapably bear the energy transition burden.[459] Therefore, the matter of energy or environmental justice

---

455 J. Stephens, *Diversifying Power: Why We Need Antiracism, Feminist Leadership on Climate and Energy* (Island Press, 2020) 24 (arguing that whether the connections between social justice and climate action is mainstreamed, there should be a de-escalation of hostility toward feminism and antiracism).

456 U. Outka, 64.

457 S. Fields, 'Continental Divide: Why Africa's Climate Change Burden is Greater?' (2005) 113 *Environmental health perspectives* 8 <https://www.ncbi.nlm.nih.gov/pmc/articles/PMC1280367/> (accessed 05 July 2020).

458 D. Shepard, Global warming: severe consequences for Africa: New report projects greater temperature increases, Africa Renewal (Dec. 2018-March 2019). <https://www.un.org/africarenewal/magazine/december-2018-march-2019/global-warming-severe-consequences-africa> (accessed 05 July 2020).

459 C.Shields, *Renewable Energy Facts and Fantasies* (Clean Energy Press 2010) 36.

is necessarily an acute problem.⁴⁶⁰ The MDGs—it is safe to say—have not arisen out profound hope when it comes to reducing social inequalities. Neither at a local nor at a global level. Consequently, the immense challenge remains for SGDs to quickly and effectively address the social catastrophe that has been unfolding for over a century.

In some African societies, there is still something of the nature of a plight and predicament. The aristocratic power structure confines a large section of the population, including indigenous people and other social groups, in an awful situation. That is the case of Cameroon, DRC, Gabon, RC, etc. In addition to the fact that most of them belong to the low-income community, people bear a disproportionate share of the economic, social and environmental burden, and is either snatched away or deprived of basic benefits stemming from any social transformation. Thereby, it seems inherent in processes such as the energy transition to cause disproportionate harm to those social groups. The example of universal access to energy services is a good starting point in this regard. Most of French-speaking African states must indeed integrate the universal access to energy services into other socio-economic problems facing them. In fact, the state is called upon to provide direction in these various problems at stake. This is also the premise that emerges from resolution 71/233. It considered the perspective of sustainable development to be the one which offers a framework conducive to the reconciliation of economic, social and environmental disparities. But also, it emphasized that access to energy for all is a core dimension of sustainable development.

With due regard to its role, the state must be able to envisage achieving energy democracy through its environmental policies aiming at ensuring universal access to energy services. The concept of energy democracy means here that energy services should be offered at affordable costs, so that it will reduce both energy and economic poverty and improve the livelihood of people. This seems to align with the vision outlined in the Congolese Strategy for the Development of the Energy Sector.⁴⁶¹ Of course, this apparent link which appears as such, derives from the formulation of the vision by itself, underpinning the vast project it carries by easily suggesting that:

---

460   Uma Outka, 64.
461   A. Nzaou-Kongo, 51.

> The vision for the development of the sector consists in making available to each Congolese citizen or any user in the urban and rural environment, sustainable energy and drinking water in sufficient quantity and acceptable quality, as well as adequate sanitation services, by making the best use of all available potential.[462]

As set out in the 2010 Decree, the Strategy for the Development of the Energy Sector clearly pursues the objective of promoting access to energy, drinking water and sanitation services for the largest possible number of people, within the framework of an environmental policy that takes into account all social strata.

After a careful review of the strategy attached to the 2010 Decree, a series of two considerations should be highlighted.

First, it seems clear that the strategy aims to integrate energy issues with social and economic issues, with a view to addressing them jointly. A key element of this approach is a constant synergy between access to energy and poverty eradication. Poverty eradication—which is also goal 1 of the 2030 Agenda—is thus referred to as a significant feature of the Congolese national strategy. It is indeed directly linked to the issue of access to energy resources. The strategy emphasizes that guaranteed access to reliable energy services should help eliminate poverty in the long term and then promote significant social integration. Accordingly, it is based primarily on the need to raise the level of access to drinking water from 75 percent in 2011 to 90 percent in 2015 in urban areas. In addition, the strategy emphasizes the importance of addressing this issue within the framework of a healthy and sustainable environment. There is no doubt that such approach tends to reconcile—at least formally—these various challenges facing the state. The Congolese strategy set the need to achieve, within the framework of state incentives, to achieve by 2015.[463]

---

462 Development Strategy for the Electric Power Decr. No. 2010-822 (n 129).

463 It is important to note that this objective is far from being achieved at the time this work is being prepared.

For the energy sector, specific objectives have been set to raise the rate of access to energy from 60% in 2011 to 90% in 2015 in urban areas, and from 25% in 2011 to 50% in 2015 in rural areas; to ensure that the twelve (12) departmental capitals and eighty-six (86) district capitals are supplied with sustainable electricity. The same applies to the drinking water sector, with a rate of access to

Such effort is encouraged by resolution 71/233, which promotes energy services as basic services.[464] The resolution attaches many virtues to guaranteeing access for all to reliable, sustainable and modern energy services at an affordable cost. For this reason, it considers that such a guarantee is conducive to a considerable reduction, if not elimination, of certain phenomena, including poverty, morbidity and mortality, and disaster risks. Thus, the reference to energy services as basic services put forth the concept of human dignity. That is to say, ensuring access to energy is a matter of human dignity, which necessarily changes life quality, but also eases access to health, education, drinking water, and sanitation. In fact, with no doubt this challenge concerns both developing and developed countries.

The quest for such synergy will certainly lead the Congolese government to point out social inequalities—in access to energy—between men and women. For sustained periods of time, the national energy policy has not considered the need to ensure access to and deployment of sustainable energy services, while improving gender equality and women's empowerment. The challenge is to raise the level of equal distribution of environmental, social and economic benefits without yielding them to privileged social groups. So, the state must address them—in a cohesive way and more often than not—as a dual challenge of achieving universal access to energy without any gender discrimination—on the one hand—and reducing women's vulnerability due to the lack of reliable energy services, on the other hand. In this sense, the synergy aligns with SDGs Agenda which encourages the state to raise awareness on and develop capacity-building programs on women in energy services. In short, the state must involve women in any national or local actions related to the transmission of knowledge and the promotion of access to energy services. Yet it does not seem incorrect to say that the Congolese strategy did not specifically emphasize the significant element. It merely encourages a global

---

    drinking water of 75% in 2011 and 90% in 2015 in urban areas, and 50% in 2011 and 75% in 2015 in rural areas; as for sanitation, the challenges are less significant, in particular access to adequate sanitation services in 56% of households in urban areas by 2015. Stratégie de développement des secteurs de l'énergie électrique, de l'eau et assainissement.

464 Developing countries are particularly targeted here, as the majority of people living on less than $1.25 a day are located in sub-Saharan Africa or Asia. These people are among the 27 million people who do not have access to clean water and who suffer from famine and malnutrition according to UNICEF.

vision of promoting energy democracy without bothering to integrate sometimes relevant dimensions or details.

One cannot fail to acknowledge that the integration of energy access and food security seems to be a serious challenge. After a relative period of economic fallout, most states in the Central African region, including Cameroon, Central African Republic, Equatorial Guinea, Gabon, and RC, have been struggling to redeem themselves from the financial meltdown underway since 2014. Soon after, they have had to scramble with the ongoing pandemic. The hard fate being almost doubled with a soaring unemployment rate. Then, the government is compelled to take a serious step when it comes to fighting hunger, improving nutrition, and guaranteeing food security. In recent years, some African states faced with food riots have seen them turn into social and political turmoil. That is, food security matters for people in those states. Indeed, universal access to energy must be promoted through a framework, which supports the intensification of agricultural production, by simplifying access to clean production technologies. Whatever the policy content contours could be, it must aim directly at the activity of small producers, including indigenous people, family farmers, livestock breeders, fishermen, and informal economy operators, etc. No doubt, this is a way to arise positive spillovers, create jobs, and add real social value.

Second, ensuring universal access to energy services must be addressed in the context of environmental policies. This is a necessary move, given the need to achieve universal access while ensuring the protection of the environment. Its ends are to meet the expectations of 75 percent of Congolese people, living mainly in rural areas, who depend on traditional biomass for cooking and heating. Similarly, is it convenient to provide a more appropriate solution to the 40 percent of Congolese people who do not have access to electricity and who, despite the availability of energy distribution services, do not benefit from them either because of the cost, or social exclusion factors.[465]

The promotion of renewable energies would thus encompass the expected impact on these social disparities. Improving widespread access to reliable, sustainable and modern energy services at an affordable cost is a

---

465 However, urban areas are not exempt from this dependence, due to customary practices.

social imperative. But if universal access is to be considered "universal," then its scope must not exclude anyone across the social strata. Such an approach should therefore harbor social acceptance of the process, and necessarily facilitate the advocacy of universal access to energy services. More specifically, to more environmentally friendly energy resources and services. In this regard, one could concur with the idea that promoting environmental protection and access to cleaner energy sources will inevitably spur people's acceptance. In other words, the perspective of sustainable development would thus be strengthened. This approach could be fundamental for the RC, which is particularly concerned about the need for transition, but is still coping with unsettling social issues, some of which are heart wrenching. This is a double challenge for the RC and any delay in the implementation of the national strategy, which does not address all these social issues, will probably haul other forms of social problems. With no doubt, it requires integrated planning and management of environmental issues and needs as part of the energy and ecological strategy.[466]

Although this is a real challenge, it must be admitted that it gives the state some leeway in the fight against climate change. Despite the apparent simplicity of such an assumption, universal access to energy would help cultivate resilience and adaptation to climate change as well as mitigation. It could also be said that this is a key element of the UN Secretary General's report on "United Nations Decade of Sustainable Energy for All."[467] The report underlines the urgent need to rethink the global energy system and ensure the provision of quality energy services to provide access to all without discrimination.[468]

In addition to reinforcing the Sustainable Energy for All initiative launched by the UN Secretary General in 2011, following the General Assembly's proclamation of 2012 as the International Year of Renewable Energy for

---

[466] In this regard, it clearly emphasizes that 'improving energy efficiency, increasing the share of renewable energy and promoting cleaner and more energy-efficient technologies are important elements of sustainable development, and that it is important to promote energy conservation, develop energy-efficient technologies and products, and establish effective mechanisms to ensure a more rational use of energy resources.'

[467] UNSG, United Nations Decade of Sustainable Energy for All: Report of the Secretary-General, U.N. Doc. A/70/422 (14 October 2015).

[468] Ibid 3.

All,[469] the report outlines the development of energy access as part of the global strategy to respond to the threat of climate change.[470] As such, the report highlights the scaling up of renewable energy deployment as a key means of doubling the capacity to adapt and build resilience to climate change. This was already reflected in a previous UN Secretary General's report, "Promotion of new and renewable sources of energy."[471] The latter took stock of the state of the global energy system and considered the broad perspective of a transition focused on three objectives: renewing the global energy mix by significantly increasing the share of renewable energy sources, achieving global energy efficiency and ensuring access to reliable and sustainable energy services for all.[472]

The General Assembly subsequently endorsed this overall dynamic, emphasizing—for its part—the need to build a modern energy system and ensure access for all to energy services as a means of resilience to climate change. Thus, paragraphs 7 and 9 of resolution 71/233 focused on this issue, encouraging states to show greater ambition and effort in this area and to report, through their nationally determined contributions, on the progressive deployment of new and renewable sources of energy. The General Assembly resolution 70/205 on "Harmony with Nature" had stated similar considerations, making direct reference to the World People's Conference on Climate Change and the Rights of Mother Earth, hosted by the Plurinational State of Bolivia in Cochabamba on April 20-22, 2010.

With the benefit of hindsight, the set of issues we have examined in this section highlights the need for African states to make room in their overall priorities for energy access issues, but also to make sense of the urgency[473] that arises from the need to interrelate energy, environmental and social issues. Consequently, African states must also consider climate change risks compromising—in view of its effects—people's universal access to energy resources and services. Such risk could create even more vulnerable groups, vulnerable to changes and their effects. The transition process

---

469 UNGA. Res. 65/151, U.N. Doc. A/RES/65/151 (20 Dec. 2010).
470 UNGA. Res. 67/215, U.N. Doc. A/RES/67/215 (21 Dec. 2012); G.A. Res. 70/201, U.N. Doc. A/RES/70/201 (22 Dec. 2015).
471 U.N. Secretary General, Promotion of new and renewable sources of energy: Report of the Secretary-General, U.N. Doc. A/69/323 (18 Aug. 2014).
472 Ibid.
473 A. Nzaou-Kongo, 51.

puts the energy system at the forefront, including through the need to modernize it, make it cleaner and safer, and make it particularly resilient to climate change. But it could be flawed if it does not reflect a real social change, especially by relying on the protection of everyone or the vulnerable. Yet, government action can only make a real difference at this time with the indispensable support of international cooperation.

4.4 Energy Transition: Between International Cooperation and Global Dynamics

State's duty to cooperate is a component of the current international peace and security realm. From international environmental to energy cooperation, of course hinged on various means, the sovereign state is a central actor.[474] Likewise, cooperation is at the heart of ensuring access to energy services. It is both the starting point of the Sustainable Energy for All initiative and an instrumental means for supporting its internal implementation. In this respect, cooperation is expected to take place within the classical framework of the UN and its specialized agencies or related institutions, but it inevitably aims to promote the transition effort over African states and be concentrated on specific issues.

4.4.1 A Contribution Within a Global Dynamic

The transition away from fossil fuels is a process. It requires a specific trajectory, a solid set of structural elements, a material implementation, and—most importantly—a diversity of actors. The overall look is about shifting from a perennial exploitation to a diversification of energy sources. Also, it is about diversification of power or leadership, horizontally and vertically.[475] The paradigm shift needed is to be fostered by a global dynamic. It suggests that the transition must also rest on international cooperation, which will necessarily support African states to move gradually—at their own pace—towards their climate targets or transition objectives. International cooperation must be the framework for moving from climate isolationism to energy democracy in the first place.[476] This means

---

[474] J. Pauwelyn, *Conflict of Norms in Public International Law: How WTO Law Relates to other Rules of International Law* (Cambridge University Press 2003) 350.

[475] J. Stephens, 51.

[476] Ibid 47.

that the changing structure of international relations is also at stake in this matter.[477] If energy transition requires a new set of ideas,[478] stemming from a diversity of sciences, including environmental science and engineering,[479] geophysics, hydrology,[480] palaeoclimatology,[481] climate science,[482] etc., then it is necessary to get it out of the infinite administrative maze, of its impoverishing fragmentation,[483] of any form of monopoly other than truly-science environmental policy.[484]

In this respect, international cooperation is proven to be of great importance. It is a key element of the Declaration on Principles of International Law concerning Friendly Relations and Cooperation among states in accordance with the Charter of the United Nations, which is deemed to set forth international society's constitutional principles.[485] Adopted by the UN General Assembly through resolution 2625 (XXV), the declaration emphasizes the obligation to cooperate in all political, economic, and social areas. As a duty of states, cooperation implies that they must work together based on bona fide motivations.[486] It is also referred to the UN as

---

477   J. Patterson, *Remaking Political Institutions: Climate Change and Beyond* (Cambridge University Press 2021) 25.

478   D. Chivers, *The No-Nonsense Guide to Climate Change: The Science, the Solutions, the Way Forward* (New Internationalist 2010) 11.

479   S. Parker, R. Corbitt, and McGraw-Hill *Encyclopedia of Environmental Science and Engineering* (McGraw-Hill 3rd ed. 1993).

480   B. Johansen, *The Encyclopedia of Global Warning Science and Technology* (ABC-CLIO 2013) 561.

481   B. Black et al., *Climate Change: An Encyclopedia of Science and History* (ABC-CLIO 2013) 1118.

482   J. Romm, *Climate Change: What Everyone Needs to Know* (Oxford University Press 2016) 1.

483   W. Boyd, Climate Change, 'Fragmentation, and the Challenges of Global Environmental Law: Elements of a Post-Copenhagen Assemblage' (2010) 32 *U. Pa. J. Int'l L.* 457.

484   J. Laitos and J. Okulski, *Why Environmental Policies Fail* (Cambridge University Press 2017) 9.

485   V. Lowe, *International Law* (Oxford Press University 2007) 90.

486   U.N. Secretary General, Report on the United Nations Decade of Sustainable Energy for All, submitted in the context of the Programme on the Promotion of New and Renewable Sources of Energy for the Implementation of Agenda 21, the Programme for the Further Implementation of Agenda 21 and the outcomes of the World Summit on Sustainable Development and the United Nations Conference on Sustainable Development. U.N. Secretary-General, United

the framework for the momentum of any cooperation dynamic. Yet when it comes to energy transition, the UN represents, moreover, the framework in which the initiative "Sustainable Energy for All" was launched. One of the ends of that initiative—which has become an international organization—is to harbor the energy democracy.[487]

Sustainable Energy for All has been the subject of numerous the General Assembly resolutions and documents.[488] Most of which promote ensuring access to energy for all, doubling the rate of improvement in energy efficiency, and doubling the share of renewable energy in the global energy mix.[489] Over that period, the UN General Assembly advocated that it is incumbent to the state to ensure universal access to energy. Accordingly, the recent resolution 71/233 appears to be the linchpin of this process. For the most part, it asserted that the UN has a prima facie duty when it comes to the global dynamic regarding climate change. The same resolution has framed different elements aiming to foster cooperation within the framework of the United Nations as regards ensuring access to energy services for all. Although the term "cooperation" is not used, it neverthe-

---

Nations Decade of Sustainable Energy for All: Report of the Secretary-General, § 13, U.N. Doc. A/72/156 (17 July 2017).

487 Sustainable Energy for All Statutes, Seforall (28 Oct. 2016) <https://www.seforall.org/system/files/2019-12/SEforALLStatutes.pdf> (accessed 05 July 2020).

488 U.N. Secretary-General, United Nations Decade of Sustainable Energy for All: Report of the Secretary-General, § 13, U.N. Doc. A/72/156 (17 July 2017); U.N. Secretary-General, International Year of Sustainable Energy for All, 2012: Report of the Secretary-General, U.N. Doc. A/67/314 (16 Aug. 2012); U.N. Secretary-General, United Nations Decade of Sustainable Energy for All: Report of the Secretary-General, U.N. Doc. A/70/422 (14 Oct. 2015); U.N. Secretary-General, United Nations Decade of Sustainable Energy for All: Report of the Secretary-General, U.N. Doc. A/68/309 (Aug. 6, 2013); U.N. Secretary-General, Sustainable Energy for All: Note by the Secretary-General, U.N. Doc. A/67/551 (02 Nov. 2012); U.N. Secretary-General, Sustainable Energy for All: A Global Action Agenda: Note by the Secretary-General, U.N. Doc. A/67/175 (Apr. 2012); UNIDO, UNIDO institutional support for the United Nations Secretary-General's initiative on sustainable energy for all: Addendum Note by the Secretariat, Doc. GC.14/18/Add.1 (Nov. 28–Dec.2, 2011); U.N. Secretary-General, Sustainable Energy for All: Note by the Secretary-General, U.N. Doc. A/66/645 (Nov. 1, 2011); G.A. Res. 65/151, U.N. Doc. A/RES/65/151 (Dec. 20, 2010).

489 UNIDO, *UNIDO institutional support for the United Nations Secretary-General's initiative on sustainable energy for all: Addendum Note by the Secretariat*, § 4, Doc. GC.14/18/Add.1 (28 Nov.–2 Dec. 2011).

less structures and shapes the initiative, both in its variations and in the possibilities of deployment of the tools for facilitating such access. Attention is also drawn to the role of "cornerstone" that the UN Secretary General and, through him, the entire World Organization must play. This role is clearly at the forefront of the resolution, when the General Assembly takes note of the report of the Secretary General on the United Nations Decade of Sustainable Energy for All, which it links to the report on the promotion of new and renewable sources of energy, but also calls for the rapid implementation of the Global Action Plan for the Decade.

UN Secretary General is invited in the first place to promote—within the Organization itself—renewable energy, energy efficiency and the adoption of sustainable practices. To this end, policy and measures adopted must be applied in all UN premises around the world. Since such action can only be effective if access to energy services is promoted at the same time, UN Secretary General must reconcile these objectives and support them in all Member States of the Organization. In addition, UN Secretary General should prepare a report on the activities carried out within the framework of the United Nations Decade of Sustainable Energy for All, in consultation with Member States and other interested parties, which should include all related activities carried out and implemented within the United Nations system.[490]

The General Assembly then invited the UN Secretary General to support the new inter-state dynamics resulting from the major challenges facing the planet. As a result, UN Secretary General has been asked to submit two reports in 2014,[491] pursuant to the General Assembly resolution

---

490 This report of the Secretary-General, together with the report on the implementation of resolution 71/233, is expected at the seventy-second session of the General Assembly, which will include a sub-item entitled 'Ensuring access to reliable, sustainable, modern and affordable energy services for all' (item 31) under the item entitled 'Sustainable development.'

491 U.N. Secretary-General, Report on the United Nations Decade of Sustainable Energy for All, submitted in the context of the Programme on the Promotion of New and Renewable Sources of Energy, envisaged for the implementation of Agenda 21, the Programme for the Further Implementation of Agenda 21 and the outcomes of the World Summit on Sustainable Development and the United Nations Conference on Sustainable Development. U.N. Secretary-General, *United Nations Decade of Sustainable Energy for All: Report of the Secretary-General*, U.N. Doc. A/69/395 (22 Sept. 2014); U.N. Secretary-General, Promotion of new and renewable sources of energy: Report of the Secretary-General, U.N.

67/215 (2012) on Promotion of new and renewable sources of energy. In many respects, a review of the two reports clearly reveals that—faced with a planet in transition—it is now more than important to reconcile the issue of new and renewable energies with that of access to these energies for as many people as possible. Moreover, it is a primary duty for UN Secretary General to make global cooperation the driving force for actions that are relayed at the local level. Indeed, the two reports give a clear account of where global multi-stakeholder partnerships are henceforth expected to be in this challenge. In this respect, the supply of energy services—which still concerns many countries—is both a priority and an emergency for the entire planet.

Therefore, UN Secretary General is compelled to making the Sustainable Energy for All initiative a global framework that can underpin actions on the ground, which will have to be financed and technically supported by the established global multi-stakeholder partnerships developed within the UN system. Indeed, it is this framework that must give considerable impetus to the promotion of renewable energy and other relevant initiatives, such as the Global Climate Action Plan. Further, the General Assembly encourages UN Secretary General to carry on his action in this area. Such action will be supported with mobilizes financial resources as well as necessary technical assistance to facilitate access to energy for all. Indeed, it is incumbent upon UN Secretary General to mobilize resources for sustainable energy supply and efficiency improvements, as well as to coordinate and fully utilize the international financial resources gathered for this purpose. These resources should enable the implementation of national and regional projects aimed at ensuring access for all to reliable, sustainable, modern and affordable energy services. UN Secretary General should broaden and increase partnerships and commitments to achieve the actions for the decade and secure new commitments for the achievement of the interim targets by 2024 and beyond. All relevant United Nations agencies are invited to participate in this initiative, so that no state—especially African states—is left behind by such mobilization.

The duty of cooperation enshrined in these various recommendations, in particular in the "Sustainable Energy for All" initiative, is indeed part of an overall dynamic in which traditional cooperation is invited to pin down

---

Doc. A/69/323 (18 Aug. 2014).

specific energy challenges, which might require countries to receive help and support from other states and actors. It is therefore a more global, but also more specific, international cooperation that must accompany the action of fossil fueled states, such as RC, towards a more appropriate energy transition.

4.4.2 The Specific Situation of Developing Countries

The new choice to be made in favor of alternative energy sources to hydrocarbons is a real challenge, with a twofold stake: first, the renewal of energy mix with new and renewable energy sources; second, ensuring universal access to energy services. Faced with this double challenge, the situation of African states requires reflection both on the forms of cooperation envisaged and on the specialized cooperation developed in Africa.[492]

First, it is appropriate to consider institutional cooperation.[493] This is envisaged through the coordination provided by UN Secretary General at the global level and aims mainly at strengthening inter-institutional and intergovernmental dialogue as well as the institutional support indispensable to meet modern energy challenges. In this regard, resolution 71/233 underpins—not without emphasis—the central role that the International Renewable Energy Agency ("IRENA") must play in assisting UN Member States to achieve their renewable energy goals.[494] For this reason, the General Assembly had expressed its appreciation for IRENA's work to promote and contribute to the widespread adoption of renewable en-

---

492 It is important to emphasize that, in addition to these forms of cooperation, there is national private action. This is still insufficiently implemented in general or in a field such as this. However, local investment could play an important role in reinforcing the bouquet with non-fossil resources and in maintaining employment. Economists are just beginning to study this type of investment, as well as that made by the diaspora to the country.

According to recent studies, more than 62 billion CFA francs are sent to Africa and could be directed towards this 'green investment.'

493 S. Furfari, 'Les agences de maîtrise de l'énergie aux niveaux régional, insulaire et urbain : l'expérience de la Communauté européenne' (1998) 38 *Liaison Énergie-Francophonie* 25.

494 Thus, the General Assembly had approved the work program and budget of this agency (point 2). Here again, it is understood that the General Assembly commits IRENA to continue its activities in the framework of energy cooperation, which is one of the reasons for its creation (Art. 4 (B) para. 4 of IRENA's Statutes).

ergy and their sustainable use. Encouraged in its energy cooperation activities, IRENA was therefore called upon to mobilize multi-stakeholder partnerships and to promote, both regionally and nationally, the supply of reliable, sustainable and modern energy services at an affordable cost in African states.

Thus, a further concern may arise from the lack of financial resources. This means that financial cooperation must be particularly emphasized. International financial institutions and donor states are invited to support the actions undertaken in the framework of the "Sustainable Energy for All" initiative. Over a long period of time, African states have built up their energy sector with a large share of fossil fuels. The paradigm shift could not be left to private actors alone, but the state must also create a dynamic. Yet financial cooperation will appear to be of great importance for the state. These initiatives should, inter alia, accompany the promotion, encouragement and accessibility of new and renewable sources of energy that are environmentally sound, climate-resilient and low-carbon. However, financing agencies are called upon to finance at reduced rates, adapted to the development structure of each developing country, the energy diversification projects submitted to them. Likewise, such financing must also raise the levels of investment in these countries in order to promote the development, deployment and wider coverage of renewable energy in their territories.

Finally, referring to the reasons set out above, scientific and technical cooperation must represent an appropriate means to transform domestic energy mixes. Fundamentally dependent on financial partnerships, this specific aspect should help to develop, disseminate, and transfer—environmentally friendly technologies—to African states by creating conditions conducive to the expected shift. At the outset, it should be noted that this dimension of cooperation seems to highlight the role the Technology Facilitation Mechanism ("TFM"), created by the Addis Ababa Programme of Action, is going to play in order to support the implementation of SDGs.[495] The goal being in fact to promote SDGs and enhance international cooperation,[496] including multi-stakeholder partnerships, civil society, private sec-

---

495 About Addis Ababa Action Agenda of the Third International Conference on Financing for Development (Addis Ababa Action Agenda). UNGA Res. 69/313, U.N. Doc. A/RES/69/313 (27 July 2015).

496 With regard to Transforming our world: the 2030 Agenda for Sustainable Devel-

tor, scientific community, UN entities and other stakeholders, etc. This cooperation should facilitate capacity building and the provision of technical assistance regarding the deployment of renewable energy.[497]

> We recognize that both public and private investment have key roles to play in infrastructure financing, including through development banks, development finance institutions and tools and mechanisms such as public-private partnerships, blended finance, which combines concessional public finance with non-concessional private finance and expertise from the public and private sector, special-purpose vehicles, non-recourse project financing, risk mitigation instruments and pooled funding structures. Blended finance instruments including public-private partnerships serve to lower investment-specific risks and incentivize additional private sector finance across key development sectors led by regional, national and subnational government policies and priorities for sustainable development. For harnessing the potential of blended finance instruments for sustainable development, careful consideration should be given to the appropriate structure and use of blended finance instruments. Projects involving blended finance, including public-private partnerships, should share risks and reward fairly, include clear accountability mechanisms and meet social and environmental standards. We will therefore build capacity to enter into public-private partnerships, including with regard to planning, contract negotiation, management, accounting and budgeting for contingent liabilities. We also commit to hold inclusive, open and transparent discussion when developing and adopting guidelines and documentation for the use of public-private partnerships and to build a knowledge base and share lessons learned through regional and global forums.

---

opment. UNGA Res. 70/1, para. 70, U.N. Doc. A/RES/70/1 (25 Sept. 2015).
497   UNGA Res. 69/313, para. 48.

Broadly, one can consider financial cooperation to increase support to research and development of appropriate technologies, to strengthen governments' actions and the private sector. These efforts will make it possible to introduce a progressive share of innovative technologies. Therefore, the General Assembly encourages financial institutions to continue the actions already undertaken and to take new initiatives to help developing countries and interested parties to plan, finance and implement sustainable energy projects, while ensuring their follow-up and the establishment of national capacities and institutions.

In light of the facts outlined above, the relevance of specialized cooperation on solar energy in Africa, referred to as the World Solar Programme 1996-2005, should be pointed out as a specific framework for cooperation on energy issues.[498] Approved—in conjunction with the Harare Declaration on Solar Energy—at the World Solar Summit held in Harare (Zimbabwe) on 16 and 17 September 1996— the World Solar Programme,[499] following the recommendations of the Rio 92 Earth Summit, aimed at improving the quality of life of populations, particularly in the rural areas of developing countries, through the use of solar energy and other energy sources.[500] Recognized as a contribution to the overall process of sustainable development by resolution 53/7, the World Solar Programme 1996-2005 was considered to ensure the continuity of the United Nations Conference on New and Renewable Sources of Energy initiative, launched in the context of the establishment of a new international economic order. The Programme had placed cooperation in favour of developing countries at the heart of its action, as shown by the numerous resolutions adopted in this regard.[501]

---

498 B. Berkovski, 'The World Solar Programme' in Renato Gavasci and Sarantuyaa Zandaryaa (eds.), *Environmental Engineering and Renewable Energy* (Elsevier Science 1998) 151.

499 The program was confirmed by the World Solar Commission in June 1997.

500 J-C. Mba-Nze, 'Vers une utilisation à grande échelle des énergies renouvelables en Afrique' (2000) 46 *Liaison Énergie-Francophonie*, 14-5.

501 For an understanding of how the United Nations has dealt with energy challenges and the development that this major issue has undergone before the General Assembly, see in this regard the resolutions as follows:

UNGA Res. 45/209, U.N. Doc. A/RES/45/209 (21 Dec. 1990); UNGA Res. 45/208, U.N. Doc. A/RES/45/208 (21 Dec. 1990); UNGA Res. 43/193, U.N. Doc. A/RES/43/193 (20 Dec. 1988); UNGA Res. 43/192, U.N. Doc. A/RES/43/192 (20 December 1988); UNGA Res. 41/170, U.N. Doc. A/

The General Assembly expressed its satisfaction with the support for the commitments already made by some donor Member States and called on all other UN Member States to contribute to the successful implementation of this program in Africa.[502] Therefore, the resolution invited the Secretary General to act in close cooperation with the United Nations Educational, Scientific and Cultural Organization ("UNESCO"), the United Nations Environment Programme ("UNEP") and other relevant organizations. On the heels of the World Solar Programme, the Programme on Promotion of New and Renewable Sources of Energy would carry on the global effort initiated and implemented 10 years earlier. Resolution 56/200 set the scope of the shift in the programme. As a matter of fact, it comprised 5 major projects, the implementation of which was envisaged only in the framework of international, regional and bilateral cooperation.[503] Afterwards, the necessity for such cooperation was iteratively recalled by 10 resolutions adopted by the General Assembly.[504] The invitation was particularly valid for South-South cooperation in which some African states, such as Morocco, have already acquired a know-how recognized worldwide. This specificity of African solar energy should not, however, prevent the evolution towards other sources of energy such as wind, which can be of great importance for coastal states, such as the

---

RES/41/170 (05 Dec. 1986); UNGA Res. 40/208, U.N. Doc. A/RES/40/208 (17 Dec. 1985); UNGA Res. 39/176, U.N. Doc. A/RES/39/176 (17 Dec. 1984); UNGA Res. 39/173, U.N. Doc. A/RES/39/173 (17 Dec. 1984); UNGA Res. 38/169, U.N. Doc. A/38/169 (19 Dec. 1983); UNGA Res. 38/151, U.N. Doc. A/RES/38/151 (19 Dec. 1983); UNGA Res. 37/251, U.N. Doc. A/RES/37/251 (21 Dec. 1982); UNGA Res. 37/250, U.N. Doc. A/RES/37/250 (21 Dec. 1982); UNGA Res. 36/193, U.N. Doc. A/RES/36/193 (17 Dec. 1981); UNGA Res. 35/204, U.N. Doc. A/RES/35/204 (16 Dec. 1980); UNGA Res. 34/190, U.N. Doc. A/RES/34/190 (18 Dec. 1979); UNGA Res. 33/148, U.N. Doc. A/RES/33/148 (20 Dec. 1978).

502 UNGA Res. 53/7, U.N. Doc. A/RES/53/7 (26 Oct. 1998).

503 These include the renewable energy education and training program, the information and communication network for Africa's renewable energy regions, rural energy production, capacity building and biomass initiatives.

504 UNGA Res. 69/225, U.N. Doc. A/RES/69/225 (19 Dec. 2014); UNGA Res. 67/215, U.N. Doc. A/RES/67/215 (21 Dec. 2012); UNGA Res. 66/206, U.N. Doc. A/RES/66/206 (22 Dec. 2011); UNGA Res. 64/206, U.N. Doc. A/RES/64/206 (21 Dec. 2009); UNGA Res. 62/197, U.N. Doc. A/RES/62/197 (19 Dec. 2007); UNGA Res. 60/199, U.N. Doc. A/RES/60/199 (22 Dec. 2005); UNGA Res. 58/210, U.N. Doc. A/RES/58/210 (23 Dec. 2003); UNGA Res. 55/205, U.N. Doc. A/RES/55/205 (20 Dec. 2000); UNGA Res. 54/215, U.N. Doc. A/RES/54/215 (22 Dec. 1999).

Congo, Angola, Cameroon.[505] Such a broadening of the energy mix can indeed enable those states to adapt to climate change and thus participate in its mitigation.

4.5 The African Union's Efforts Towards the States

The role of the African Union is illustrative of the efforts undertaken by international organizations of this nature to support the action of states in the energy transition process.[506] In this regard, I propose to focus on a few initiatives that are under discussion or that have been proposed by the specialized technical committees ("STC")[507] to the Commission of the African Union ("CAU").

It is important to note that one of the efforts of the African Union that strengthens the role of the state in the energy transition is the African Union Commission's Initiative on the Harmonized Continental Regulatory Framework for the Electricity Sector.[508] The governance of this initiative, which is still in its infancy, will be entrusted to a steering structure comprising a board of directors composed of five Ministers for Energy,[509] each representing a region of the continent, as well as a technical steering committee led by the heads of the power pools under the Continental Transmission Master Plan.[510]

---

505 With the prospect of marine wind farms. B. Saulnier, 'L'homme et l'énergie éolienne' (1997) 35 *Liaison Énergie-Francophonie* 4; Ch. Falvin, 'L'énergie éolienne : un vent nouveau qui souffle de plus en plus fort' (1997) 35 *Liaison Énergie-Francophonie* 17.

506 A. Nzaou-Kongo, 'African Union' (2020) 31 *Yearbook of International Environmental Law* 1.

507 STC on Transport, Transcontinental and Interregional Infrastructure, Energy, and Tourism.

508 Strategy for the Development of a Harmonised Regulatory Framework for the Electricity Market in Africa, African Union (June 18, 2021), <https://au.int/sites/default/files/documents/40438-doc-Strategy_HarmonisedRegulatoryFrameworkElectricityMarket.pdf> (accessed 05 July 2020).; Action Plan for Harmonized Regulatory Framework for the Electricity Market in Africa, African Union (June 18, 2021), <https://au.int/en/documents/20210618/action-plan-harmonised-regulatory-framework-electricity-market-africa> (accessed 05 July 2020).

509 A. Nzaou-Kongo, 'African Union' (2019) 30 *Yearbook of International Environmental Law* 8.

510 B. Odetayo and M.Walsh, 'A policy perspective for an integrated regional power

For the implementation of the pilot phase of this initiative, the STC on Trade, Industry and Minerals ("STC-TTIIET") had adopted—during its second ordinary session held in Cairo, Egypt, on April, 16–17, 2019—the "Guidelines for Minimum Energy Performance Standards ("MEPS") and Energy Labeling," which will be implemented by Member States.[511] These measures aim to strengthen energy markets at the regional and continental level. However, this effort can only be effectively sustained if support is given to certain countries where access to energy is not assured, and where the need to set up mini-electricity distribution networks in both urban and rural areas is still pressing.[512] It is certain that the contribution of a harmonized electricity market will only be significant to the extent that a number of social justice issues are resolved, such as health, tourism, education and access to information, etc.[513]

In line with the ambition to improve energy performance, the African Energy Efficiency Program is another African Union initiative,[514] which aims to promote energy transition. Here again, the role of the state is crucial. Its overall objective is to significantly reduce final energy demand and polluting emissions in Africa. In addition to this, the program also aims to ensure broad access to electricity, competitiveness, energy security and economic development.[515] This program will be supported by the African Energy Sector Transition Initiative, which will initiate and promote "Deep Decarbonization Pathways ("DDP")" in regions and selected pilot countries in Africa. Its aim is to provide states with the necessary guidance on transforming their energy systems in the short, medium, and long term.[516]

Another initiative, which is undoubtedly competing with the first one in terms of the role of the state and public actors, is the African Renewable

---

pool within the Africa Continental Free Trade Area' (2021) 156 *Energy Policy*.

511 AUC/STC-TTIET, *Guidelines for Minimum Energy Performance Standards (MEPS), Energy Efficiency, Energy Labeling and Eco-Design at the Continental Level*, AU Doc. IEA24387 (April 14-18, 2019).

512 A. Nzaou-Kongo, (n 510) 8.

513 Ibid.

514 African Union, Energy Efficiency, (2021), <https://au-afrec.org/energyefficiency.php> (accessed 05 July 2020).

515 A. Nzaou-Kongo (n 510) 11.

516 Ibid 12.

Energy Initiative ("AREI").[517] Its purpose is to support access to energy services and the migration to low-carbon development by mobilizing public and private investment to finance energy infrastructure. This initiative will concretely facilitate the training of national actors and develop the necessary measures to encourage various initiatives and projects aimed at renewable energies in the Member States. In this sense, the AREI Independent Delivery Unit (IDU) will submit projects to the AREI Board of Directors for adoption and implementation. Similarly, the IDU will be able to promote knowledge sharing in renewable energy research.[518]

However, this effort aimed at access to sustainable energy is now conceived only in the framework of the achievement of the United Nations Sustainable Development Goal 7 ("SDG7"). Thus, the SE4ALL organization is led to work with the African Union to promote universal access to sustainable energy by 2030. In this respect, various actions will be carried out concerning the initiative "electricity for all in Africa," which will make it possible to release the financing of the electricity projects.

The growing importance of these issues has led the Commission to work on the implementation of a policy framework and guidelines on bioenergy in Africa. In this sense, a bioenergy development strategy and investment plan for the East and Central African regions was adopted by the STC-TTIIET. In this regard, the CTS

> "(...) requested AUDA-NEPAD to assist Member States in mapping bioenergy resources and identifying priority bioenergy projects for development and implementation and to request development partners to provide the necessary support."

In the same sense, it had encouraged the development of strategies for other regions. Consequently, the Bioenergy Program aims to give materiality to this policy framework. In particular, it aims to build capacity to improve monitoring, reporting and sustainability of bioenergy in the countries of the continent. It will enable states to have the capacity to col-

---

517 African Union, Africa Renewable Energy Initiative, (2016), <http://www.arei.org/wp-content/uploads/2018/09/AREI-Framework.pdf> (accessed 05 July 2020).
518 A. Nzaou-Kongo, 9.

lect and analyze data on bioenergy and especially to put in place a solid system of continuous monitoring through the implementation of sustainability indicators for bioenergy ("GSI"). These data will make it possible to know the exact figures of bioenergy production and consumption in Africa.[519]

In order to achieve the shift towards more sustainable, reliable and affordable energy systems, the transition process will also benefit from efforts that will be undertaken within the framework of the African Clean Energy Corridor ("ACEC").[520] This is a regional initiative with a dual purpose. On the one hand, to encourage states and stakeholders to develop and move towards clean, indigenous and cost-effective renewable energy; on the other hand, to promote cross-border trade in renewable energy. This effort was initially implemented in the Eastern and Southern African power pool region, notably the Africa Clean Energy Corridor under the IRENA.[521] This effort had been extended to the region of energy pools of West Africa in 2016 under the name of West Africa Clean Energy Corridor ("WACEC"). In this context, it is expected that the cooperation of states with international organizations, projects of interconnection of energy networks can promote the deployment of renewable energy and that technical and technological knowledge transfers can follow.[522]

These cooperation efforts to support state action also concern African island states. This is particularly true of the program for the development of solar energy and small hydroelectricity in Africa, which aims to encourage the development of renewable energy in African island states.[523] This program is initiated within the framework of the African Union Commission to promote the development of solar energy, small hydro and renewable energy in these states that have specific problems and must also participate in the energy transition process. In addition to focusing on the contribution of solar and small hydro in their new energy mix, the program is intended to encourage African island states to gradually re-

---

519   Ibid 11.
520   Ibid 10.
521   Ibid.
522   Ibid.
523   Ibid 7.

duce their dependence on fossil fuels.[524] It is now recognized that these states are facing extreme manifestations of climate change and that the climate risk to which they are exposed is becoming increasingly acute; for this reason, the program envisages placing special emphasis on resilience to these phenomena and encouraging these states to use the renewable energy resources at their disposal. In this way, the Commission will be able to assist African island states to access financing for renewable energy deployment, they can make use of financing obtained through climate efforts or resort to other sources of financing.[525]

4.6 Conclusion

The state remains an important player in the new energy system. This is so even if its central place in the international system, which has long been undisputed, is in fact diminishing. As a sovereign actor, it is nevertheless necessary to recall that the state remains the driving force behind most of the major decisions taken by its institutional components and intended to be applied on its territory. This fact is accompanied by the participation of states in international organisations at various levels, in which they defend their national positions and interests. The legal mechanisms for incorporating into the domestic legal order the decisions or legal instruments adopted in these fora fall within the competence of the state. This inevitably underlines the role of the state in coordinating efforts related to the energy transition at domestic, regional and international levels. This role seems to be important, because beyond that, it is the responsibility of the state to mobilise the funds and support necessary for the implementation of its energy and climate policy.

Therefore, many other actors come into play, including foreign investors, multinational, national and local companies, private groups and even individuals. At this level, while many of these actors, whose list is not exhaustive, are known to be the recipients of norms and policies, they are also actors who participate in the formation and implementation of laws and regulations, where mechanisms for consultation and dialogue are provided. These are moreover social dynamics to which the state is now subject. The issue of energy transition shows how difficult it becomes for

---

524 Ibid.
525 Ibid.

the state, for example when it manages to implement its climate policy without consultation, to escape a contentious review of its policy when non-state actors point out the inadequacy between the climate ambition displayed in the policy and the international commitments it has also made. This is an important lever that strengthens climate democracy, without often calling into question the discretionary power of a government which, because of the technical and scientific nature of the issues it deals with, must make informed and reasoned decisions.

Although their activist position is *per se* the reason for a form of antagonism towards the state, non-state actors have a much more complementary and necessary role to play, as the recent development of climate litigation shows. It is towards this path of democratic pluralism that the chances of success for the transformation of the energy system are now moving. However, this will only be possible if no external contingency hinders the trajectory that certain states are setting themselves in terms of energy transition with the support of all the players. Indeed, nothing is guaranteed, as a political changeover or ideological rivalries are enough to break or reverse the trajectory of certain states in this area. Non-state actors therefore literally have a duty of care over the progress made, a kind of anti-return ratchet, to ensure that whatever changes occur, the initial trajectory is eventually suspended, but never reversed. In this respect, it must be understood that energy transition as a social transformation is a challenge for present and future generations.

Bibliography

1. Vaclav Smil, Energy: a Beginner's Guide 127 (Oneworld Publications 2006).

2. Vaclav Smil, Energy and Civilization: A History 295 (The MIT Press 2017).

3. Vaclav Smil, The Earth's Biosphere: Evolution, Dynamics, and Change 231 (The MIT Press 2002).

4. Lavanya Rajamani & Jacqueline Peel (eds.), The Oxford Handbook of International Environmental Law 1 (Oxford University Press 2ed. 2021).

5. Felwine Sarr, Afrotopia 5 (Drew S. Burk & Sarah Jones-Boardman trans. University of Minnesota Press 2019).

6. Thomas Paine, Common Sense 05 (Penguin Books-Great Ideas ed. 1776).

7. Max Weber, Le Savant et le Politique 22-86 (Union Générale d'Éditions ed. 1963).

8. Francis Fukuyama, The End of History and the Last Man 16 (The Free Press ed. 1992).

9. Leon Duguit, Traite de Droit Constitutionnel 534 (E. de Boccard 3d ed. 1927).

10. Max Weber, L'Ethique Protestante et l'Esprit du Capitalisme 140 (1st ed. 1905).

11. Emile Durkheim, Montesquieu et Rousseau: Precurseurs de la Sociologie 115-198 (1st ed. 1918).

12. Raymond Carre de Malberg, Contribution À La Théorie Générale de l'État: Spécialement d'Après Les Données Fournies Par Le Droit Constitutionnel Français 11 (Librairie du Recueil Sirey 1st ed. 1920).

13. Andrew Hurrell, On Global Order: Power, Values, and the Constitution on International Society 6 (Oxford University Press 1st ed. 2008); David Lewis and Nazneen Kanji, Non-Governmental Organizations and Development 5 (Routledge ed. 2009).

14. Beham Markus P., State Interest and the Sources of International Law. Doctrine, Morality, and Non-Treaty Law 251 (London: Routledge 2018).

15. Rosemary Lyster & Adrian Bradbrook, Energy Law and the Environment 78 (Cambridge University Press 2006).

16. Jennie C. Stephens, Elizabeth J. Wilson, and Tarla Rai Peterson, Smart Grid (R)Evolution: Electric Power Struggles 405 (Cambridge University Press 2015).

17. Jennie C. Stephens, Diversifying Power: Why We Need Antiracism, Feminist Leadership on Climate and Energy, 24 (Island Press 2020).

18. Joost Pauwelyn, Conflict of Norms in Public International Law: How WTO Law Relates to other Rules of International Law 350 (Cambridge University Press 2003).

19. James J. Patterson, Remaking Political Institutions: Climate Change and Beyond 25 (Cambridge University Press 2021).

20. Danny Chivers, The No-Nonsense Guide to Climate Change: The Science, the Solutions, the Way Forward 11-12 (New Internationalist 2010).

21. Sybil P. Parker & Robert A. Corbitt, McGraw-Hill Encyclopedia of Environmental Science and Engineering (McGraw-Hill 3rd ed. 1993).

22. Bruce E. Johansen, The Encyclopedia of Global Warning Science and Technology 561 (ABC-CLIO 2013).

23. Brian C. Black et al., Climate Change: An Encyclopedia of Science and History 1118 (ABC-CLIO 2013).

24. Josephe Romm, Climate Change: What Everyone Needs to Know 1 (Oxford university Press 2016).

25. Jan Laitos & Juliana Okulski, Why Environmental Policies Fail 9 (Cambridge University Press 2017).

Epilogue

Victor B. Flatt, Alfonso López de la Osa Escribano and Aubin Nzaou-Kongo

Important perspectives emerge from the current evolution of energy law and policy. There is no doubt that the diversity of actors, urging for the development of the discipline in phase with the current stakes, plays an important role. The same applies to the increasingly important role of case law in both energy and climate matters. Whatever the stated ambition of the laws on energy transition, energy or climate, the discipline now necessarily involves integrating into the analysis highly refined scientific complexities, but also jurisprudential developments that allow to feel the pulse of various considerations: ambition, realistic or unrealistic efforts, enforceability, legal certainty, etc.

On the one hand, the caselaw underlines the importance of energy transition or the share of renewable energy in the mix. Only recently, in a decision dated 13 March 2020 outlining the judge's reasons for not ruling, it was specifically noted that "renewable energy ... is of great public interest and concern." In this recent case, in which the applicant sought leave from the judge to challenge a decision of the UK Secretary of State for Business, Energy and Industrial Strategy, dated 1 May 2019. This decision favored applications for UK support for renewable energy generation made by electricity operators in the context of the operation of offshore or remote island wind turbines, while it did not take into account generators using onshore non-remote island wind turbines. In this respect, the question as it was brought before the judge was whether: European competition law prevents the UK government from honouring manifest commitments by political parties to reduce or eliminate public subsidies to onshore wind farms? The parties, having reached a compromise while the court proceedings were underway, asked the judge not to hand down judgment, but rather to hear them in a private hearing. The judge argued that the renewable energy issue was a matter of public interest and required the utmost attention and therefore should be heard in a public hearing.

> However, what is more important, in my view, is that this is a public law application for public law remedies, against a public authority on a subject, renewable energy, that is of very significant public interest and concern. Judicial review actions are neither private law disputes

between individuals nor private law disputes between a citizen and a state agency. Rather, they are public law disputes which must, except in the most unusual circumstances, be heard and decided in public. Normally, no party or combination of parties should be able to prevent the publication of a court decision after the proceedings have concluded.

On the one hand, the British government was concerned about the inevitable impact that a court case would have on the U.K.'s efforts to achieve the ambitious target of zero net carbon emissions by 2050. It was felt that the proceedings would create uncertainty and doubt about the validity of contracts awarded under the Round 3 ("AR3") allocation of state support agreements, and more importantly, the unpredicted impact that the proceedings could have on the significant investment that the UK is promoting in renewable energy.

> ... The energy projects in question are said to involve billions of pounds of investment. These investments could be further delayed, perhaps significantly, while these proceedings are ongoing, even if the end result is not to invalidate the relevant CFD contracts. It is said that, beyond the immediate commercial consequences for actual businesses and third parties, this continued delay and uncertainty has significant implications for the UK's plans to achieve its net zero carbon emissions target.

The UK Secretary of State had relied on a public interest, including the maintenance of legal certainty and the proper functioning of the AR3 contracts regime. In granting this request, the judge therefore concluded that:

> However, it seems to me that there is an overriding factor here, namely that a withdrawal of this application by consent, and in the absence of a judgment being given, gives the parties, including the Secretary of State, legal certainty in relation to AR3. This, in my view, serves to improve the prospects for the development and deployment of renewable energy in this country, and puts the UK in a better position to deal with climate change. This is a factor of overwhelming public importance, out-

weighing the commercial interests of individual participants, and even the public interest in accessing a judgment in a case like this. This consideration leads me to grant this request.

On the other hand, non-state actors continue to press through litigation, directly targeting energy or climate issues, for energy and climate law to have a dual materiality, to be adapted to current issues and to lead to the responsibility of the state and other private actors.

A topical example is the judgment of the Supreme Court of Ireland of 31 July 2020. The contribution of this case when it comes to climate issues is very significant. It underlines that climate change is one of the greatest concerns facing mankind. As a result, climate change issues must be addressed, and predominantly by governments, because states are in the first instance responsible of environment protection on their territory. Hence, it seems no longer enough for states to tackle climate change anyhow, including designing climate policies that are far from the realities on the ground, scientific realm or failing to vindicate the rights guaranteed by pertinent legal instruments.

The leave to appeal sought by the Friends of the Irish Environment (FIE) was in line with this approach. The facts, as acquired at trial, can thus be summarized as follows: pursuant to Article 3 of the Climate Action and Low Carbon Development Act (the "2015 Act"), the Government of Ireland published on the 19 July 2017, the National Mitigation Plan. The object of the plan was to provide a climate action policy framework for initiating and pursuing the transition to a low carbon, climate resilient and environmentally sustainable economy. First, the plan pointed out that the Irish climate action aimed at contributing to the global response to the climate challenge, which is based on the United Nations Framework Convention on Climate Change, the Kyoto Protocol and the Paris Agreement. In this regard, the plan pledges to pursue the objectives of restricting global temperature rise, and endeavour the collective effort to limit the temperature increase to 1.5°C. Then, the plan referred to the European Union (EU) action as to the objectives of reducing greenhouse gas emissions, and the contribution of the Republic of Ireland to the EU effort, supporting at the same the Irish approach of fostering climate resilience and low greenhouse gas emissions development.

In this context, FIE, a non-governmental organisation, actively involved in protection and promotion of the environment, instituted proceedings before Justice MacGrath. First, because the state failed to approve acceptable means to reduce greenhouse gas emissions in compliance with the objectives of the UNFCCC, the Kyoto Protocol and the Paris Agreement, the applicant contended that the Plan was unconstitutional. Accordingly, the Plan would violate the rights of the applicant, its members, and the public, namely the rights enshrined in the Charter of Fundamental Rights and Freedoms. Second, FIE argued that the plan was adopted in breach of the European Convention on Fundamental Rights. Finally, the applicant claimed that the plan was ultra vires the powers of the Minister under the 2015 Act. To sustain its contention, the applicant alleged that the decision of the Irish Government to approve the Plan should be abrogated, because it did not comply with the section 4 of the 2015 Act. It was also argued that the Plan did not contain any appropriate measures susceptible "to achieve the management of a reduction of greenhouse gas emissions in order to attain emission levels appropriate for furthering the achievement of the National Transition Objective."

FIE raised the issue that the government plan would fail to achieve the desired objectives and the necessary steps for a better evaluation of the government action on mitigating concentrations throughout the period during which the reduction of emissions was made and supported. It was the responsibility of the government to prevent the irreversible damage that could occur from maintaining the current level of GHG emissions in the atmosphere. FIE also emphasised the necessity of reducing the emissions in a steep direction and in line with the projection need to support international regulation relating to climate issues. It argued that despite the effort of combating climate change, the planned trajectory of emissions and the level of concentration would still alter the atmosphere. As regards Government Plan, its objective of achieving a reduction by 2050 was sustained by mechanisms that do not prevent the risk of harm inherent to the short-term incautious reduction of emissions.

The applicant broadly emphasised that scientific evidence suggested that net negative carbon dioxide emissions should be maintain within 2°C. In this sense, it appeared indispensable that political efforts should be made in favour of scenarios that allow the development of carbon dioxide removal technologies such as bioenergy, extensive reforestation or forest growth. Accordingly, FIE raised the point that the mitigation plan should be based on concrete calculations that can define a new trajectory of emission reductions in the short term in a serious and substantial way. To support this, it was argued that the government's focus on long-term targets would simply affect carbon budgets. In this regard, FIE contended that by developing and approving the plan, the government was not doing enough to combat climate change. As a result, the plan was approved in breach of law and of the applicant's rights, which it was for the courts to review and uphold.

After considering the arguments and conclusions reached by the parties, the judge held that, in light of the discretion exercised by the government in developing and adopting the Plan under the 2015 Act, it could not be decided that the government had breached the provisions of the Act in the manner alleged. Accordingly, it was concluded that the Plan referred to the state's obligations under European Union law and existing international agreements. In this regard, it could not be considered to contain any proposal to achieve the national objective of transition which, as provided for in Article 3(1), would lead to the transition by 2050. Moreover, the court noted that the Plan did indeed specify the policy measures that the government considers necessary to reduce greenhouse gas emissions. These reasons led the court to dismiss the FIE proceedings. Therefore, FIE sought leave to appeal to the Court.

FIE was granted leave to appeal by decision of 13 February 2020. This leave was substantial by the fact that there was a common understanding in the arguments of both applicant and defendant on the need for urgent action to combat climate change. It also appeared that the parties did not dispute the importance of understanding and putting in place the means to combat climate change, in particular with regards to the likely increase in greenhouse gas emissions during the duration of the plan. In this regard, the leave was granted on the basis of the Parties' common understanding of the seriousness of the likely effects of climate change.

In this case, the judge came to some interesting conclusions.

Firstly, the ruling focused on the standing of FIE. As it is a corporate body not enjoying in itself the right to life or the right to bodily integrity, the judgement concluded that FIE does not have standing in respect of the rights it seeks to assert under the Constitution or the ECHR. In support of this, the judgement found that it was not necessary to grant FIE standing under the exception to the general rule.

Second, the judgement highlighted numerous considerations with respect to the merits of the case. At the outset, it was stressed that FIE should be allowed to make a broad range of arguments in support of its written submissions. Similarly, the justiciable nature of the issues raised did not allow for a finding of inadmissible encroachment by the courts into areas of policy, as the policy aspects of the issue were framed by the 2015 Act.

It was also held that the 2015 Act, by virtue of its Article 4, provided for a sufficient level of specificity in the measures identified in the plan with a view to achieving the national transition objective by 2050, which alone made it possible to make a reasonable assessment of the realistic nature of the plan and the policy options it contained. The judgement therefore recalled that the adoption of the 2015 Act has led to the use of certain mechanisms such as public participation in the process leading to the adoption of the plans and transparency regarding the government's official policy on the national transition objective. In this regard, it was noted that the plan advocated by the government should be sufficiently precise, even if it was not certain that the details of the evolution of the climate situation were known. To this extent, the plan should be able to define the policy to be followed over the entire period up to 2050. Consequently, it would not be sufficient to set up a five-year plan.

In view of these considerations, the judgement concluded that the plan should be quashed, as it "... [fell] far short of the level of specificity required to ensure such transparency and to comply with the provisions of the 2015 Law." The judgement strongly emphasized that a plan of such a nature could not be sustained in the future. These new trends will inevitably enrich energy law and policy but will also help rethink the way the discipline has been taught to date. This poses a new challenge to the teaching of energy law, since it is now necessary to draw all the consequences of the climate litigation that is developing before us.

Index

Aarhus Convention on Access to Information, Public Participation in Decision-making and Access to Justice in Environmental Matters, 76

Abstentions, 102

Access to energy, 96, 97, 102, 112, 113, 114, 115, 117, 118, 119, 120, 121, 122, 123, 125, 126, 127, 133, 134

Addis Ababa Programme of Action, 129

Administrative Court of Paris, 18

Administrative power, 80, 81, 82, 99, 112

Adriatic Sea, 63,

Aegean-Levantine Sea, 63

Affordable energy, 25, 110, 114, 126, 127, 135

Affordable cost, 112, 117, 199, 120, 128

African Clean Energy Corridor: ACEC, 135, 136

African Union Commission, 133, 135, 136

African Energy Efficiency Program, 134,

African Renewable Energy Initiative: AREI, 134,

African states, 95, 96, 97, 99, 101, 102, 113, 114, 116, 117, 120, 122, 123, 127, 128, 129, 132, 134

African Union, 99, 132, 133, 134, 135, 136

Ambitious climate goals, 105,

Anthropocene Era, 95,

Arctic, 32, 46, 48, 49, 50, 51, 52, 53, 59, 89

Arctic region, 50, 51

Artic Offshore Oil and Gas Guidelines, 52, 53, 61, 75, 109

Atlantic, 31, 45, 48, 54, 56, 59

Authorizations, 67, 78, 102,

Availability, 90, 95, 110, 120,

Baltic, 27, 46, 48, 54, 56, 57, 58, 83, 89

Baltic Sea Region, 58

Barents Sea, 46

Best available technology, 58

Best environmental practice, 58

Binding international standards, 51

Binding Target, 13, 15, 18, 19, 20,

Bioenergy Program, 135,

Biofuel, 2, 3, 4, 8, 11, 14, 16, 107,

Biofuels, 2, 3, 4, 8, 11, 14, 16

Biomass, 4, 10, 11, 107, 108, 111, 120, 132,

Black Sea, 46, 48, 49, 50, 53, 54, 55, 56, 83, 89

Blue Economy, 47, 90

Blue Growth, 47, 48, 80, 84, 85,

90, 91

Brexit, 45

Bucharest Convention on the Protection of the Black Sea Against Pollution, 54

Capacity to adapt and build resilience, 121

Carbon capture and storage, 6

Central Mediterranean Sea, 63

Changing structure of international relations, 123

Civil society, 100, 101, 107, 129

Clean cooking fuels, 96

Clean Energy for all Europeans, 3, 7,

Climate change, 6, 7, 8, 24, 32, 45, 50, 52, 82, 90, 91, 98, 115, 116, 121, 122, 123, 124, 125, 132, 136, 142, 143, 144, 145,

Climate isolationism, 123

Climate neutrality, 6

Climate objectives, 8, 10,

Climate plans, 19

Climate policies, 109, 143

Climate policy, 5, 6, 7, 45, 109, 137,

Climate Policy Choices, 6

Coercive power, 100

Collective Obligation, 18, 19

Common future of mankind, 95

Common good, 98

Competence, 4, 5, 7, 13, 15, 16, 24, 29, 30, 31, 78, 79, 109, 137

Competences, 4, 5, 7, 13, 15, 16, 24, 29, 30, 31, 78, 79

Conference of the Parties, 6

Conservation, 54, 76, 77, 82, 83, 88, 97

Constitution 124, 144, 145

Constitution of Angola, 102

Constitutional law, 26, 99, 100, 105

Constitutional Law Rationale, 100

Consumption, 1, 4, 5, 6, 7, 8, 9, 10, 11, 12, 13, 14, 15,16, 18, 19, 20, 21, 24, 25, 26, 45, 96, 97, 135

Consumption of RE, 5, 8, 14

Continental shelf, 55, 61, 62, 63, 64, 65, 66, 68, 69, 70, 71, 72, 73, 74, 75, 83, 88, 89

Convention for the Protection of the Mediterranean Sea against Pollution, 62

Convention on the Protection of the Marine Environment of the Baltic Sea Area, 54, 57

Cooking technologies, 96

Cooperation, 62, 64, 73, 75, 79, 80, 81, 99, 109, 122, 123, 124, 125, 126, 127, 128, 129, 130, 131, 132, 136

Court of Justice of the European Union: CJEU, 1, 9, 16, 17, 18

COVID-19, 23, 119

Customs union, 30

Decarbonisation, 28, 34, 38, 105,

106, 134

Deep Decarbonization Pathways: DDP, 134

Deepwater Horizon platform, 46, 67, 68

Dependable sources, 95, 97, 106, 107, 108, 110, 120, 121, 122, 125, 129, 132,

Development Agenda beyond 2015, 113

Development of renewable energies, 11

Directive (EU) 2018/2001, 1, 2, 3, 4, 5, 6, 8, 10, 11, 13, 14, 15, 16, 18, 19, 20, 21

Directive 2001/77/EC, 2, 8, 11, 15

Directive 2003/30/EC, 2, 8, 11, 12, 15, 16

Directive 2008/56/EC, 53, 59, 63, 70, 77, 83, 84

Directive 2009/28/EC, 2, 3, 7, 11, 15, 16, 18, 20

Directive 2013/30/EU, 50, 51, 70, 76, 79

Directive on safety of offshore, 51, 79, 80

Disaster response, 47

Discretionary power, 99, 105, 109, 137

Disharmonious way of life, 98

Disproportionate distribution of power and wealth, 98

Domestic legal order, 102, 137

Domestic legal system, 102

Domestic needs, 97

Drilling, 46, 52, 66, 77

Ecological damage, 18

Economic impacts, 96

Economic landscape, 96

Economy, 23, 34, 37, 40, 46, 47, 88, 90, 110, 120, 143,

ECT, 24, 31, 32, 33, 34, 38, 40

ECT modernisation, 40

Effective Transition Effort, 112

Efficiency, 36, 39, 95, 97, 114, 121, 122, 125, 127, 133, 134

Electricity, 2, 3, 11, 14, 15, 19, 35, 105, 106, 107, 108, 110,115, 118, 120, 122, 133, 134, 135, 136, 141

Emissions: GHG, 1, 2, 4, 6, 7, 8, 9, 18, 32, 78,134, 142, 143, 144, 145

EnC framework, 34, 35, 36

EnC objectives, 35

EnC Treaty

Energy Community Treaty, 24, 31, 34, 36, 41,

Energy, 1, 2, 3, 4, 5, 6, 7, 8, 10, 11, 12, 13, 14, 15, 16, 17, 18, 19, 20, 21, 22, 23, 24, 25, 26, 27, 28, 29, 30, 31, 32, 33, 34, 35, 36, 37, 38, 39, 40, 45, 46, 49, 51, 55, 70, 81, 82, 83, 84, 89, 90, 95, 96, 97, 98, 99, 100, 102, 104, 105, 106, 107, 108, 109, 110, 112, 113, 114, 121, 123, 125, 127, 129,

131, 135, 137, 141, 143, 144, 145

Energy and Climate Union, 20

Energy Charter Conference, 20

Energy Charter Treaty: ECT, 24, 31, 32, 33, 34, 38, 40

Energy choices, 5

Energy Community, 23, 24, 31, 34, 35, 36, 38, 40, 45, 46, 47

Energy consumption, 1, 4, 5, 7, 9, 14, 16, 18, 19, 21, 24, 25, 26, 45

Energy demand, 25, 26, 27, 37, 39, 40, 110, 134

Energy democracy, 117, 119, 123, 124

Energy law approach, 25, 34, 99, 105, 112, 115

Energy market, 2, 3, 14, 16, 27, 28, 30, 31, 34, 35, 38, 39, 88, 89, 133, 134

Energy mix, 4, 10, 11, 12, 13, 16, 27, 28, 29, 30, 32, 38, 39, 40, 95, 97, 108, 109, 110, 113, 114, 122, 125, 127, 129, 132, 136

Energy policy, 5, 7, 18, 28, 29, 32, 36, 38, 39, 51, 95, 99, 107, 109, 110, 111, 112, 115, 119, 133

Energy resources, 3, 30, 38, 39, 97, 108, 110, 118, 120, 121, 122, 135, 136

Energy Sector, 7, 23, 27, 31, 33, 34, 37, 39, 96, 105, 107, 109, 110, 111, 113, 114, 117, 118, 128, 134

Energy Sector Transition Initiative, 134

Energy security, 8, 12, 13, 14, 23, 25, 26, 27, 28, 29, 30, 32, 34, 35, 36, 37, 38, 39, 40, 41, 50, 51, 65, 66, 67, 75, 79, 90, 101, 106, 107, 119, 120, 122, 134

Energy services, 96, 97, 110, 112, 113, 114, 115, 117, 118, 119, 120, 121, 122, 125, 126, 127, 128, 134

Energy supply, 8, 12, 13, 23, 24, 26, 27, 28, 34, 37, 39, 40, 51, 106, 107, 111, 126, 127, 128

Energy transit, 32

Energy transition, 2, 25, 96, 97, 98, 99, 109, 112, 116, 117, 122, 123, 124, 127, 132, 133, 134, 136, 137, 138, 141

English-speaking countries, 97, 99, 105

Environment, Public Health and Food Safety: ENVI, 81

Environmental and climate ordeal, 96

Environmental challenge, 56, 64

Environmental justice, 76, 77, 99, 112, 115, 116, 134, 144

Environmental plight, 98

Environmental policies, 74, 117, 120, 124, 140

Environmental protection, 7, 8, 33, 34, 52, 53, 65, 72, 86, 90, 120

Environmentally friendly

technologies, 113, 129
Environmentally sound sources, 95, 110, 129
Environmentally sustainable manner, 106, 107
Espoo Convention on Environmental Impact Assessment in a Transboundary Context, 76
EU dependency, 45
EU energy diplomacy, 28
EU external relations, 23, 27
EU law, 3, 13, 17, 18, 21, 36, 69, 70, 71, 73, 87,
EURATOM Treaty, 12
European Commission, 26, 28, 33, 35, 38, 50, 54
European Community, 6, 57, 62
European Council, 7, 13, 28
European Economic Area: EEA, 45, 31
European External Action Service, 29
European governance mechanism, 20
European Legal Order, 73, 76
European Maritime Safety Agency: EMSA, 78, 79, 85
European Neighbourhood Policy: ENP, 54
European objectives, 7
European Parliament, 2, 5, 7, 9, 10, 20, 34, 35, 36, 45, 47, 48, 49, 50, 51, 52, 53, 54, 56, 58, 59, 60, 63, 67, 68, 69, 70, 76, 77, 78, 79, 80, 81, 83, 88

European target, 14, 19

European Union, 1, 4, 6, 9, 10, 12, 23, 25, 38, 40, 45, 46, 47, 48, 49, 50, 51, 53, 54, 56, 57, 58, 60, 62, 63, 64, 65, 67, 68, 69, 70, 71, 72, 73, 78, 79, 81, 85, 87, 88, 89, 90, 91, 92, 105, 143, 145

European waters, 46, 50

Europeanisation, 1, 2, 3, 4, 11, 12, 13, 14, 15, 18, 19, 20, 21

Eurostat, 16, 21, 24, 25, 26, 27, 37, 40, 45, 49

Executive branch, 103, 109

Exploitation, 12, 32, 46, 47, 50, 51, 55, 57, 58, 61, 62, 63, 64, 65, 66, 68, 69, 70, 71, 72, 73, 75, 76, 79, 80, 81, 82, 83, 84, 89, 91, 106, 115, 123

Exploitation phase, 58

Exploration, 46, 47, 52, 55, 57, 58, 61, 62, 63, 64, 65, 66, 68, 69, 71, 72, 73, 75, 78, 79, 80, 81, 82, 83, 84, 89, 105

Exploration phase, 58

Feed-In-Tariff, 108

Floating, 58, 66

Formal soft law, 49, 60, 66, 85, 109

Fossil fuel civilization, 95

Fossil fuel exporters, 23

Fossil fuel industries, 33

Fossil fuel subsidies, 115

Fossil fuels, 7, 25, 26, 27, 32, 34, 37, 95, 96, 111, 112

Framework of the Barcelona System, 61, 62

French-speaking countries, 99, 102, 117

Fuel suppliers, 14

Functions of the state, 99, 101

Future development, 13, 95

Future need, 95

Gambia's Renewable Energy Act, 108

GDP, 37

General Assembly, 90, 112, 113, 115, 121, 122, 124, 125, 126, 127, 128, 130, 131, 132,

Geothermal, 11, 107

GHG Emissions, 1, 6, 7, 9, 32, 144

Global catastrophe on nature and the ecosystem, 98

Global strategy, 121

Governance, 20, 21, 25, 34, 45, 49, 50, 51, 52, 54, 55, 61, 62, 63, 64, 65, 101

Green Deal, 33

Greenhouse gas

GHG, 1, 2, 4, 6, 7, 8, 18, 32, 143, 144, 145

Guidelines for Minimum Energy Performance Standards: MEPS, 133

Guiding Principles and Strategic Directions for Offshore Oil and Gas Activities, 60

Gulf of Mexico disaster, 80

Hard Law Framework, 49, 52, 61

Harmonized Continental Regulatory Framework for the Electricity Sector, 133

Harmony with nature, 98, 122

Healthy and sustainable environment, 118

HELCOM Commission, 58

Helsinki Convention, 56, 57, 58

Human dignity, 119

Humanity, 95

Hydro, 11, 15, 107, 136

Hydrocarbons, 24, 45, 50, 51, 70, 78, 127

IA: International agreements, 7, 31, 54, 73, 145

Independent Delivery Unit: IDU, 134

Indigenous people, 98, 99, 116, 120

Industrial and economic domination, 98

Industrial civilization, 95

Industrial Revolution, 96

Industry Research and Energy: ITRE, 81

Institutional capacities, 113

Integrated Maritime Policy, 47, 48, 49, 51, 56, 58, 63, 70, 73, 77, 84, 85, 90

Index

Integrated Maritime Policy Action Plan, 51

Integration, 7, 12, 27, 45, 49, 53, 54, 65, 70, 71, 92, 118, 119

Internal market, 14, 30, 31, 39

International agreements, 7, 31, 54, 73, 145

International cooperation, 47, 49, 79, 99, 122, 123, 124, 127, 129

International Court of Justice: ICJ, 55

International Energy Agency: IEA, 31

International Renewable Energy Agency: IRENA, 32, 96, 128, 136,

International Solar Alliance, 31, 32

International Trade Supply, 26

International Year of Renewable Energy for All, 121

Investment in renewables, 23

Ionian Sea, 63, 64

Judicial review, 102, 141

Jurisdictional function, 101, 104

Jurisdictional power, 101

Kyoto Protocol, 6, 8, 35, 143, 144

Landfill gas, 11, 107

Lands, 6, 25, 59, 62, 87, 89, 98

Laws, 19, 34, 55, 74, 77, 80, 86, 101, 102, 103, 104, 106, 109, 112, 137, 141,

Legislative function, 101

Legislative power, 102, 104

Legislative process, 101, 102

Lisbon Treaty, 28, 30

Living environment, 98

Low-carbon techniques, 115

Marine environment, 50, 52, 53, 54, 55, 57, 59, 60, 61, 62, 63, 70, 74, 75, 77, 83, 84, 85

Marine environmental policy, 53, 59, 63, 70, 77

Marine Strategy Framework, 53, 59, 63, 69, 70, 77, 84

Marine Strategy Framework Directive, 53, 59, 63, 69, 70, 77, 84

Material law, 49, 60, 66, 78, 80, 82

Material soft law, 49, 60, 66, 78, 80, 82

Mauritius Renewable Energy Agency: MARENA, 108, 109

Mauritius Renewable Energy Agency Act, 108, 109

Mediterranean, 46, 48, 49, 61, 62, 63, 64, 65, 66, 67, 68, 70, 71, 73, 74, 75, 79, 80, 81, 82, 84, 86

Mediterranean Sea, 54, 63, 64, 68, 69, 71, 72, 83, 89

Member States, 3, 4, 6, 7, 10, 11, 12, 13, 14, 15, 16, 17, 18, 19, 20, 21, 26, 27, 28, 29, 30, 31, 34, 35, 38, 39, 40, 46, 48, 49, 50, 52, 53, 60, 63, 67, 68, 69, 70, 71, 72, 73, 77, 79, 80, 81, 83, 85, 86, 87, 88, 89, 105, 126, 128, 131, 133,

134, 135

Methane hydrates, 66

Millennium Development Goals: MDGs, 113

Mix, 4, 10, 11, 12, 13, 16, 27, 28, 29, 30, 32, 38, 39, 40, 95, 97, 108, 109, 110, 113, 114, 122, 125, 127, 129, 132, 136

Modern energy, 90, 110, 112, 113, 118, 120, 122, 128

Monopoly of legitimate violence, 100

Moratorium, 52

National Assembly, 102

National Energy Mix, 4, 10, 11, 12, 13, 16, 109, 114

National objectives, 18, 19, 20

National security, 29, 37

National target, 13, 15, 16, 20, 108

Nationally determined contributions, 6, 122

Natural gas share, 25

Natural harmony, 96

Natural resources, 25, 27, 51, 55, 74, 77, 98, 109

New Energy System, 95, 96, 105, 136

Non-binding provisions, 60

Non-state actors, 97, 99, 101, 137, 138, 143

North Sea, 45, 59

Norwegian Sea, 46

Ocean energy, 11, 107

Oceans, 47, 55, 73, 84, 90

Offshore, 47, 48, 49, 50, 51, 52, 53, 55, 56, 58, 60, 61, 64, 65, 66, 67, 68, 69, 70, 71, 72, 73, 75, 76, 78, 79, 80, 81, 82, 83, 84, 85, 86, 87, 88, 89, 90, 141

Offshore Activities, 46, 49, 52, 56, 58, 61, 64, 82, 87, 88, 92, 93

Offshore Protocol, 61, 64, 68, 69, 71, 72, 73, 79, 82, 92

Oil and gas, 46, 47, 48, 50, 51, 52, 53, 55, 58, 60, 61, 64, 65, 66, 68, 75, 76, 78, 80, 81, 82, 83, 84, 85, 86, 87, 88, 89, 91, 110

Oil and gas production, 46, 61

Oil and gas reserves, 45, 50

Ordinary legislative, 5

OSPAR Area, 56, 58, 59

Paradigm shift, 95, 99, 123, 128

Paris Agreement, 6, 7, 8, 31, 32, 33, 143, 144

Player, 7, 95, 107, 108, 136, 137,

Policy choice, 5, 6

Policy Objectives, 105, 109

Political or an economic union, 23

Portuguese-speaking countries, 101

Prevention of Pollution from Offshore Activities, 58

Primary law, 1, 7

Primary services, 95

Principle of conferral, 1

Prohibitions, 102
Promise of development, 95, 96, 98
Promise of economic development, 98
Promotion of Renewable Energies, 11, 120
Proportionality, 4, 30
Protection of the environment, 7, 8, 33, 34, 52, 53, 65, 72, 86, 90, 120
Protocols, 54, 62, 63, 64
Public power, 101
Public Utilities Regulatory Authority, 108
Public International Law, 49, 67, 75, 89, 123,
RE consumption, 4, 5, 6, 8, 9, 11, 12, 13, 14, 15, 16, 18, 19, 20,
RE production, 14
Reduction of CO_2 emissions, 2, 6, 7, 9, 12, 32, 45, 119, 144, 145
Regalian missions, 100
Regional Center for Emergency Response against Accidental Marine Pollution: REMPEC, 65
Regional Systems, 49, 56, 89
Regional Treaties, 49, 56
Regulation, 19, 20, 21, 34, 35, 36, 48, 49, 55, 56, 58, 60, 61, 68, 70, 73, 74, 75, 76, 77, 78, 79, 80, 81, 82, 85, 86, 88, 89, 103, 104, 105, 106, 109, 137, 144

Regulatory power, 103, 104
Reliable energy, 118, 119
Renewable energy, 1, 2, 3, 4, 5, 8, 9, 10, 11, 13, 14, 16, 17, 21, 23, 24, 25, 26, 32, 33, 35, 95, 97, 106, 107, 108, 109, 112, 114, 115, 116, 121, 122, 125, 127, 128, 129, 131, 132, 134, 135, 136, 141, 142
Renewable fuels, 2, 3, 8, 11, 14, 16
Renewable gas, 15
Renewable sources: REs, 2, 3, 4, 5, 7, 8, 11, 14, 15, 17, 18, 19, 121, 122, 124, 125, 126, 129, 131, 132
Renewables, 1, 2, 3, 4, 5, 8, 9, 10, 11, 13, 14, 16, 17, 21, 23, 24, 25, 26, 32, 33, 35, 95, 97, 106, 107, 108, 109, 112, 114, 115, 116, 121, 122, 125, 127, 128, 129, 131, 132, 134, 135, 136, 141, 142
Renewables companies, 23
Renewable future, 23, 96
Republic of Congo: RC, 113, 116, 119, 120, 127
REs: Renewable energies, 1, 2, 3, 4, 5, 8, 9, 10, 11, 13, 14, 16, 17, 21, 23, 24, 25, 26, 32, 33, 35, 95, 97, 106, 107, 108, 109, 112, 114, 115, 116, 121, 122, 125, 127, 128, 129, 131, 132, 134, 135, 136, 141, 142
Resources, 25, 27, 30, 38, 39, 50, 51, 55, 66, 70, 73, 74, 77, 84, 88, 89, 90, 91, 97, 98, 108, 109,

110, 113, 115, 118, 120, 121, 122, 127, 128, 135, 136

Responsible Network Utility, 108

Right to development, 95

Rule of law, 34, 101

Safe sources, 8, 34, 46, 47, 50, 51, 52, 65, 67, 68, 73, 77, 78, 79, 80, 81, 82, 83, 84, 85, 95, 98, 106, 108, 110, 115, 116, 122

Safety and environmental credentials, 46

Safety standards, 52, 81

Seabed, 57, 63, 64, 65, 66, 68, 69, 71, 72, 74, 75

Seas, 46, 47, 53, 55, 59, 61, 62, 64, 65, 84, 88, 89, 90, 92,

Sectoral Target, 13

Sewage gas, 107

Social, 8, 40, 45, 47, 48, 49, 56, 58, 60, 81, 84, 96, 97, 98, 99, 100, 101, 107, 108, 111, 112, 114, 115, 116, 117, 118, 119, 120, 121, 122, 124, 130, 134, 137, 138

Environmental and climate burden 116, 117,

Economic burden, 96, 97, 98,

Social burden, 96, 97, 98

Soft Law Framework, 49, 60, 66, 78, 80, 85, 109

Solar, 11, 24, 31, 32, 130, 131, 132, 136

Sovereign power, 99

Sovereign state, 23, 123

Sovereignty, 12, 13, 48, 72, 83, 88, 89, 102

Specialized technical committees: STC, 133

State, 1, 3, 4, 5, 6, 7, 8, 9, 10, 11, 12, 13, 14, 15, 16, 17, 18, 19, 20, 21, 22, 23, 24, 25, 26, 27, 28, 29, 30, 31, 32, 33, 34, 35, 36, 37, 38, 39, 40, 41, 42, 43, 44, 45, 46, 47, 48, 49, 50, 51, 52, 53, 54, 55, 56, 57, 58, 59, 60, 61, 62, 63, 64, 65, 66, 67, 68, 69, 70 71, 72, 73, 74, 75, 76, 77, 78, 79, 80, 81, 82, 83, 84, 85, 86, 87, 88, 89, 90, 91, 92, 93, 94, 95, 96, 97, 98, 99, 100, 101, 102, 103, 104, 105, 106, 107, 108, 109, 110, 111, 112, 113, 114, 115, 116, 117, 118, 119, 120, 121, 122, 123, 124, 125, 126, 127, 128, 129, 131, 132, 133, 134, 135, 136, 137, 138, 139, 141, 143, 144, 145

State functions, 101

State missions, 100, 101, 134

Statutory Arrangements, 105

STC on Trade, Industry and Minerals, 133

Steering the Transition, 112

Strategy for the Development of the Congolese Energy Sectors, 113

Subsidiarity, 4, 30

Subsoil, 57, 58, 63, 64, 65, 66, 68, 69, 71, 72

Index

Supranational level, 30, 32

Sustainability indicators for bioenergy: GSI, 135

Sustainable development, 8, 32, 46, 51, 75, 90, 91, 98, 99, 109, 110, 113, 115, 116, 117, 120, 121, 124, 126, 129, 130, 131, 135

Sustainable development goals: SDGs, 32, 90, 109, 113, 135

Sustainable energy, 25, 40, 114, 117, 119, 121, 122, 123, 124, 125, 126, 127, 128, 130, 135

Sustainable Energy for All initiative, 121, 123, 134, 125, 126, 127, 128

Sustainable exploitation, 47

Technology Facilitation Mechanism: TFM, 129,

TFEU, 1, 2, 3, 4, 5, 6, 7, 8, 9, 10, 11, 12, 13, 17, 22, 29, 30, 31, 48

The 1992 Convention on the Protection of the Black Sea against Pollution, 54

The 1994 Partnership and Cooperation Agreement, 38

The 2010 Decree, 113, 117, 118

The 2011 RE Act in Ghana, 106

The Framework Directive, 53

Trade relations, 23, 24, 26, 30, 37

Transboundary dimension, 48, 84

Transport, 2, 3, 4, 8, 9, 10, 11, 14, 17, 23, 25, 34, 40, 66, 82, 84, 115, 133

Treaty in investor-state dispute settlement: ISDS, 32

Treaty of Lisbon, 2, 29, 31

Treaty on European Union: TEU, 1, 4, 12, 29

Umbrella-treaty, 64, 66

UN General Assembly, 90, 124, 125

UN Secretary General, 121, 125, 126, 127, 128

UNCLOS, 73

United Nations: UN, 6, 8, 32, 55, 56, 59, 69, 73, 76, 83, 90, 112, 113, 121, 124, 125, 126, 127, 131, 132, 135, 143

United Nations Convention on the Law of the Sea, 55, 73, 83, 90

United Nations Environment Programme: UNEP, 62, 132

United Nations Framework Convention on Climate Change: UNFCCC, 6, 8, 143

Universal access to energy, 113, 117, 119, 120, 121, 122, 124, 125, 127

Warsaw Process, 33

West Africa Clean Energy Corridor: WACEC, 136

Western Mediterranean Sea, 63

Wind, 1, 11, 17, 24, 107, 132, 141

World People's Conference on Climate Change and the Rights of Mother Earth, 122

World Solar Programme 1996-2005, 130, 131, 132

World Solar Summit, 130, 131, 132

World Trade Organisation: WTO, 32

Zimbabwe's National Energy Policy, 110, 111

Related Titles from Westphalia Press

The Zelensky Method
by Grant Farred

Locating Russian's war within a global context, The Zelensky Method is unsparing in its critique of those nations, who have refused to condemn Russia's invasion and are doing everything they can to prevent economic sanctions from being imposed on the Kremlin.

China & Europe: The Turning Point
by David Baverez

In creating five fictitious conversations between Xi Jinping and five European experts, David Baverez, who lives and works in Hong Kong, offers up a totally new vision of the relationship between China and Europe.

Masonic Myths and Legends
by Pierre Mollier

Freemasonry is one of the few organizations whose teaching method is still based on symbols. It presents these symbols by inserting them into legends that are told to its members in initiation ceremonies. But its history itself has also given rise to a whole mythology.

Resistance: Reflections on Survival, Hope and Love
Poetry by William Morris, Photography by Jackie Malden

Resistance is a book of poems with photographs or a book of photographs with poems depending on your perspective. The book is comprised of three sections titled respectively: On Survival, On Hope, and On Love.

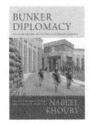

Bunker Diplomacy: An Arab-American in the U.S. Foreign Service
by Nabeel Khoury

After twenty-five years in the Foreign Service, Dr. Nabeel A. Khoury retired from the U.S. Department of State in 2013 with the rank of Minister Counselor. In his last overseas posting, Khoury served as deputy chief of mission at the U.S. embassy in Yemen (2004-2007).

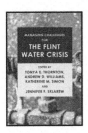

Managing Challenges for the Flint Water Crisis
Edited by Toyna E. Thornton, Andrew D. Williams, Katherine M. Simon, Jennifer F. Sklarew

This edited volume examines several public management and intergovernmental failures, with particular attention on social, political, and financial impacts. Understanding disaster meaning, even causality, is essential to the problem-solving process.

Donald J. Trump, The 45th U.S. Presidency and Beyond International Perspectives
Editors: John Dixon and Max J. Skidmore

The reality is that throughout Trump's presidency, there was a clearly perceptible decline of his—and America's—global standing, which accelerated as an upshot of his mishandling of both the Corvid-19 pandemic and his 2020 presidential election loss.

Brought to Light: The Mysterious George Washington Masonic Cave
by Jason Williams, MD

The George Washington Masonic Cave near Charles Town, West Virginia, contains a signature carving of George Washington dated 1748. Although this inscription appears authentic, it has yet to be verified by historical accounts or scientific inquiry.

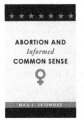

Abortion and Informed Common Sense
by Max J. Skidmore

The controversy over a woman's "right to choose," as opposed to the numerous "rights" that abortion opponents decide should be assumed to exist for "unborn children," has always struck me as incomplete. Two missing elements of the argument seems obvious, yet they remain almost completely overlooked.

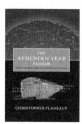

The Athenian Year Primer: Attic Time-Reckoning and the Julian Calendar
by Christopher Planeaux

The ability to translate ancient Athenian calendar references into precise Julian-Gregorian dates will not only assist Ancient Historians and Classicists to date numerous historical events with much greater accuracy but also aid epigraphists in the restorations of numerous Attic inscriptions.

The Politics of Fiscal Responsibility: A Comparative Perspective
by Tonya E. Thornton and F. Stevens Redburn

Fiscal policy challenges following the Great Recession forced members of the Organisation for Economic Co-operation and Development (OECD) to implement a set of economic policies to manage public debt.

Growing Inequality: Bridging Complex Systems, Population Health, and Health Disparities
Editors: George A. Kaplan, Ana V. Diez Roux, Carl P. Simon, and Sandro Galea

Why is America's health is poorer than the health of other wealthy countries and why health inequities persist despite our efforts? In this book, researchers report on groundbreaking insights to simulate how these determinants come together to produce levels of population health and disparities and test new solutions.

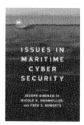

Issues in Maritime Cyber Security
Edited by Dr. Joe DiRenzo III, Dr. Nicole K. Drumhiller, and Dr. Fred S. Roberts

The complexity of making MTS safe from cyber attack is daunting and the need for all stakeholders in both government (at all levels) and private industry to be involved in cyber security is more significant than ever as the use of the MTS continues to grow.

A Radical In The East
by S. Brent Morris, PhD

The papers presented here represent over twenty-five years of publications by S. Brent Morris. They explore his many questions about Freemasonry, usually dealing with origins of the Craft. A complex organization with a lengthy pedigree like Freemasonry has many basic foundational questions waiting to be answered, and that's what this book does: answers questions.

Contests of Initiative: Countering China's Gray Zone Strategy in the East and South China Seas
by Dr. Raymond Kuo

China is engaged in a widespread assertion of sovereignty in the South and East China Seas. It employs a "gray zone" strategy: using coercive but sub-conventional military power to drive off challengers and prevent escalation, while simultaneously seizing territory and asserting maritime control.

Frontline Diplomacy: A Memoir of a Foreign Service Officer in the Middle East
by William A. Rugh

In short vignettes, this book describes how American diplomats working in the Middle East dealt with a variety of challenges over the last decades of the 20th century. Each of the vignettes concludes with an insight about diplomatic practice derived from the experience.

Anti-Poverty Measures in America: Scientism and Other Obstacles
Editors, Max J. Skidmore and Biko Koenig

Anti-Poverty Measures in America brings together a remarkable collection of essays dealing with the inhibiting effects of scientism, an over-dependence on scientific methodology that is prevalent in the social sciences, and other obstacles to anti-poverty legislation.

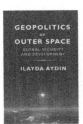

Geopolitics of Outer Space: Global Security and Development
by Ilayda Aydin

A desire for increased security and rapid development is driving nation-states to engage in an intensifying competition for the unique assets of space. This book analyses the Chinese-American space discourse from the lenses of international relations theory, history and political psychology to explore these questions.

westphaliapress.org

Policy Studies Organization

The Policy Studies Organization (PSO) is a publisher of academic journals and book series, sponsor of conferences, and producer of programs.

Policy Studies Organization publishes dozens of journals on a range of topics, such as European Policy Analysis, Journal of Elder Studies, Indian Politics & Polity, Journal of Critical Infrastructure Policy, and Popular Culture Review.

Additionally, Policy Studies Organization hosts numerous conferences. These conferences include the Middle East Dialogue, Space Education and Strategic Applications Conference, International Criminology Conference, Dupont Summit on Science, Technology and Environmental Policy, World Conference on Fraternalism, Freemasonry and History, and the Internet Policy & Politics Conference.

For more information on these projects, access videos of past events, and upcoming events, please visit us at:

www.ipsonet.org

Made in the USA
Middletown, DE
05 December 2022

15988758R00106